Also by Alan Lightman

Searching for Stars on
an Island in Maine

Searching for Stars on an Island in Maine

Alan Lightman

Pantheon Books, New York

All rights reserved. Published in the United States by Pantheon Books,
a division of Penguin Random House LLC, New York, and distributed
in Canada by Random House of Canada, a division of
Penguin Random House Canada Limited, Toronto.

Pantheon Books and colophon are registered trademarks
of Penguin Random House LLC.

Library of Congress Cataloging-in-Publication Data
Name: Lightman, Alan P., [date]
Title: Searching for stars on an island in Maine / Alan Lightman.
Description: New York : Pantheon Books, 2018
Identifiers: LCCN 2017014750. ISBN 9781101871867
(hardcover : alk. paper). ISBN 9781101871874 (ebook).
Subject: LCSH: Cosmology—Miscellanea.
Classification: LCC QB981.L545 2017. DDC 523.1—dc23.
LC record available at lccn.loc.gov/2017014750

www.pantheonbooks.com

Jacket photograph by Galaxy IC 342: T. A. Rector/University of Alaska,
Anchorage; H. Schweiker/WIYN; and NOAO/AURA/NSF
Jacket design by Kelly Blair

Printed in the United States of America
First Edition
2 4 6 8 9 7 5 3 1

Dedicated to
Rabbi Micah Greenstein and Venerable Yos Hut Khemacaro

Contents

Searching for Stars on
an Island in Maine

Cave

1979. Smell of damp earth and stone. In the dim light, a small group of people talk in hushed voices as if entering a church, spellbound by the paintings on the rock wall: bison and mammoth and horse, colored with red ochre made from dirt and charcoal and bound with saliva and animal fat. I am without words, another ghost in this primordial cave in southwestern France. Font-de-Gaume it is called. The cave paintings date to 17,000 BC and were discovered by a local schoolmaster a century ago. Hand-drawn shapes swerve and flow following the natural contours of the stone walls. In one image, a fat horse bends down as if nuzzling a bison with bent head. Elsewhere, a herd of horses gallop across the stone plains, including a large animal with orange torso and black neck, and smaller beasts speckled in black and white. One painting in particular steals my attention—an entire bison drawn from what appears to be a single flowing line.

Clearly, these early humans were consummate artists with a heightened connection to nature. Did they also believe in an ethereal world? Did they believe in the invisible? What did they think of thunder and lightning, wind, stars overhead, their own beginnings and ends? They rarely lived past the age of thirty. Clad in the skins of animals they had killed and aware of their own impending demise, they must have looked up toward the unchanging stars with awe, and desire. In the foothills beyond the caves, these ancient people buried their dead in sewn garments and surrounded the prone bodies with tools and food for the next life. Was this time and this place where the longing began?

Nearby, someone strikes a match, against regulations, and we all turn in surprise to watch the small fire. Shadows shift on the walls. Then the flame is gone, like these primitive ancestors of eons ago, like all living things, like the material world.

Longing for Absolutes
in a Relative World

For many years my wife and I have spent our summers on an island in Maine. It's a small island, only about thirty acres in size, and there are no bridges or ferries connecting it to the mainland. Consequently, each of the six families who live on the island has their own boat. Some of us were not nautical people at first, but over the years we have all learned by necessity. Most challenging are trips to the island at night, when the landmasses are only dim shapes in the distance and you must rely on compass headings or faint beacons to avoid crashing into rocks or losing your way. Nevertheless, some of us do attempt the crossing at night.

My story concerns a particular summer night, in the wee hours, when I had just rounded the south end of the island and was carefully motoring toward my dock. No one was out on the water but me. It was a moonless night, and quiet. The only sound I could hear was the soft churning of the engine of my boat. Far from the distracting lights of the mainland, the sky vibrated with stars. Taking a chance, I turned off my running lights, and it got even darker. Then I turned off my engine. I lay down in the boat and looked up. A very dark night sky seen from the ocean is a mystical experience. After a few minutes, my world had dissolved into that star-littered sky. The boat disappeared. My body disappeared. And I found myself falling into infinity. A feeling came over me I'd not experienced before. Perhaps a sensation experienced by the ancients at Font-de-Gaume. I felt an overwhelming connection to the stars, as if I were part of them. And the vast expanse of time—extending from the far distant past long before I was born and then into the far distant future long after I will die—seemed compressed to a dot. I felt connected not only to the stars but to all of nature, and to the entire cosmos. I felt a merging with something far larger than myself, a grand and eternal unity, a hint of something absolute. After a time, I sat up and started the engine again. I had no idea how long I'd been lying there looking up.

I have worked as a physicist for many years, and I have always held a purely scientific view of the world. By that, I mean that the universe is made of material and nothing more, that the universe is governed exclusively by a small number of fundamental forces and laws, and that all composite things in the world, including humans and stars, eventually disintegrate and return to their component parts. Even at the age of twelve or thirteen, I was impressed by the logic and materiality of the world. I built my own laboratory and stocked it with test tubes and petri dishes, Bunsen burners, resistors and capacitors, coils of electrical wire. Among other projects, I began making pendulums by tying a fishing weight to the end of a string. I'd read in *Popular Science* or some similar magazine that the time for a pendulum to make a complete swing was proportional to the square root of the length of the string. With the help of a stopwatch and ruler, I verified this wonderful law. Logic and pattern. Cause and effect. As far as I could tell, everything was subject to numerical analysis and quantitative test. I saw no reason to believe in God, or in any other unprovable hypotheses.

Yet after my experience in that boat many years later, I understood what Lord Indra of the Vedas must have felt when he first drank soma and could see the light of the gods. I understood the powerful allure of the Absolutes—ethereal things that are all-encompassing, unchangeable, eternal, sacred. At the same time, and perhaps paradoxically, I remained a scientist. I remained committed to the material world.

Every culture in every era of human existence has had some concept of Absolutes. Indeed, one might group a large number of notions and entities under the heading of Absolutes: absolute truth (valid in all circumstances), absolute goodness, constancies of various kinds, certainties, cosmic unity, immutable laws of nature, indestructible substances, permanence, eternity, the immortal soul, God. A philosopher friend tells me that for him and other philosophers, the word "Absolute" means "ultimate reality." I want to use the word in a different manner. Ultimate reality, whatever that is, may or may not include indestructible substances, absolute truths, omniscient beings, and the rest.

Clearly, the concepts I have labeled as Absolutes are not identical. Some are concerned with substance, others with intangible essence or abstract ideas. But they share the qualities of permanence and changelessness, ubiquity, perfection. All refer to an enduring and fixed reference point that can anchor and guide us through our temporary lives. Although the Absolutes are often associated with religion, none of them, except for the immortal soul and for God, are necessarily religious ideas. Some people would call them spiritual ideas. Many of the Absolutes are rooted in personal experience, but they involve beliefs beyond that experience.

A fascinating feature of the Absolutes—in fact, a defining feature—is that there is no way to get there from here, that is, from within the physical world. There is no *gradual*, step-by-step path to go from relative truth to absolute truth, or to go from a long period of time to eternity, or from limited wisdom to the infinite wisdom of God. The infinite is not merely a lot more of the finite. Indeed, the *unattainability* of the Absolutes may be part of their allure.

Finally, the tenets of the Absolutes have not been proven, nor can they be proven, certainly not in the way that science has proven the existence of atoms or the law of the pendulum swing. *Unprovability* is a central feature of all Absolutes. Yet I did not need any proof of what I felt during that summer night in Maine looking up at the sky. It was a purely personal experience, and its validity and power resided in the experience itself. Science knows what it knows from experiment with the external world. Belief in the Absolutes comes from internal experience, or sometimes from received teachings and culture-granted authority.

The Absolutes comfort us. Imperfect beings that we are, we can imagine perfection. In search of meaning and how best to live our lives, we can turn to irrefutable precepts and principles. Certain of our material death, we can find solace in the permanence of our ethereal souls—or, for the early humans in Font-de-Gaume, in continued hunting and living in some other life. Plato discussed absolute justice in *The Republic*. Aristotle made all terrestrial material out of earth, water, fire, and air—but he reserved for the heavenly bodies a fifth element called the *aether*, indestructible and divine. The Buddha taught the Four Noble Truths in his first sermon. Saint Augustine ascribed absolute truth to God. Newton found a proper scaffolding for his universal laws of motion in his belief in absolute space, which "in its own nature without relation to anything external, remains always similar and immovable." The unattainable Absolutes might also be considered our ultimate strivings, the best and most beautiful we can imagine. As poet C. F. Cavafy wrote, "The loveliest music is the [music] that cannot be played."

Throughout history, various Absolutes have been linked to objects in the physical world. Ancient Egyptian stone texts, discovered in the pyramid of Unas, suggested that the dead pharaohs would enter heaven through two bright northern stars known as *kihemusek,* "the indestructibles." Evidently, the perceived permanence of the stars was linked to human immortality. Plato was less elitist. He chose the stars to be the final destination of *all* moral human beings after their fleeting sojourn on earth: "And having made [the universe], the Creator divided the whole mixture into souls equal in number to the stars, and assigned each soul to a star . . . He who lived well during his appointed time was to return and dwell in his native star." For the ancient Egyptians and Greeks and many subsequent cultures, heaven, God, immortality, and physical stars were all connected to each other.

Another example is the atom, the tiny and indestructible unit of matter first proposed by the ancient Greeks. *Primordia* the atoms were called. Or *atomos.* The hypothesized atoms could not be divided into smaller components. As the smallest elements of physical matter, atoms helped unify everything else. Atoms represented indestructibility, indivisibility, unity. Obviously, atoms could not be seen with ancient Greek technology. But they were thought to exist in material form.

Belief in various Absolutes is alive and well in the world today. A new survey of 35,000 adults by the Pew Research Center found that 89 percent of Americans believe in God, and 74 percent believe in life after death—that is, in some form of immortality. A somewhat older survey by the Barna Group, an organization devoted to religion and culture, found that 50 percent of Christians in America believe in some form of absolute truth, while 25 percent of non-Christians do so. Buddhists worldwide believe in the Four Noble Truths. Hindus worship Brahman, the embodiment of eternal and absolute truth. Belief in certain physical manifestations of the Absolutes is also alive and well. A 2014 Gallup survey found that 42 percent of Americans believe in the constancy of species—in particular, that humans were created in their present form in the first days of the planet.

In the last couple of centuries and especially in recent decades, many of the Absolutes have been challenged by discoveries in science. Nothing in the physical world seems to be constant or permanent. Stars burn out. Atoms disintegrate. Species evolve. Motion is relative. Even other universes might exist, many without life. Unity has given way to multiplicity. I say that the Absolutes have been challenged rather than disproved, because the notions of the Absolutes cannot be disproved any more than they can be proved. The Absolutes are ideals, entities, beliefs in things that lie beyond the physical world. Some may be true and some false, but the truth or falsity cannot be proven.

As for science, its domain is restricted to the physical world. Science does not have any particular authority for beliefs that lie beyond that world, such as God, the soul, or notions of absolute truth and absolute goodness. Yet to the extent that various of the Absolutes have been associated with aspects of the physical world, they have been brought into question by science. The new scientific evidence accords with other findings by anthropologists and sociologists suggesting that absolutes do not exist in human societies. From all the physical and sociological evidence, the world appears to run not on absolutes but on relatives, context, change, impermanence, and multiplicity. Nothing is fixed. All is in flux.

In aesthetics and culture, we have long been accustomed to the absence of absolute standards—that is, we more or less accept aesthetic and cultural *relativism,* the dependence on context. It is in the physical sciences that the Absolutes have met their strongest trial. Let me briefly sketch a bit of the evidence here, with more to come in later chapters.

One might begin in the seventeenth century, with the reluctant acceptance that the earth is not fixed and motionless as it seems but spins on its axis and orbits the sun. That fact was proven in the 1850s by observing the slow rotation of the plane of a swinging pendulum. You can see such a pendulum, called a Foucault pendulum, at the National Museum of American History in Washington, D.C. There, a 240-pound brass bob swings from a steel cable hung two floors above. As its plane of movement slowly rotates around, the pendulum bob knocks down red pegs in a circle. According to the laws of physics, the plane of the pendulum remains fixed in space. So it must be the earth that is rotating, not the pendulum.

Moving from the earth to the sky, advances in astronomy have shown that the once eternal stars, home of departed pharaohs, eventually exhaust their limited nuclear fuel and burn out. And to the small: scientists in the late nineteenth and early twentieth century discovered that the once indivisible atom could be splintered, to reveal smaller pieces within. About the same time, Einstein's relativity demolished the appealing idea of a condition of absolute rest. Another disturbing thesis of Einstein's relativity: even the passage of time is not absolute as it seems, but depends on the relative motion of clocks.

Finally, the cosmological discoveries. In the late 1920s, astronomers discovered that the universe as a whole is not the grand and unchanging cathedral believed for millennia on end. Instead, the cosmos is expanding and stretching like a giant balloon being inflated, with all the galaxies hurtling away from each other. According to the best modern science, our universe began as a nugget of ultra-high density some 14 billion years in the past. And most recently, physicists have suggested that even our entire universe may not be a singular totality, a unity, but instead only one of a vast number of universes, called the "multiverse," each with different properties and many possibly without life.

On the one hand, such an onslaught of discovery presents a cause for celebration. In fact, the wonders of Einstein's relativity and the idea of the Big Bang were the engines that propelled me into science decades ago. Is it not a testament to our minds that we little human beings with our limited sensory apparatus and brief lifespans, stuck on our one planet in space, have been able to uncover so much of the workings of nature? On the other hand, we have found no physical evidence for the Absolutes. And just the opposite. All of the new findings suggest that we live in a world of multiplicities, relativities, change, and impermanence. In the physical realm, nothing persists. Nothing lasts. Nothing is indivisible. Even the subatomic particles found in the twentieth century are now thought to be made of even smaller "strings" of energy, in a continuing regression of subatomic Russian dolls. Nothing is a whole. Nothing is indestructible. Nothing is still. If the physical world were a novel, with the business of examining evil and good, it would not have the clear lines of Dickens but the shadowy ambiguities of Dostoevsky.

The categories of the Absolutes and what I will call the Relatives—the relativity and impermanence and multiplicity found by modern science—do not neatly separate nonscientists from scientists. Individual people may commit to some elements of the Absolutes and to some of the Relatives, and I have defined these categories so broadly that we would expect that to be the case. For example, a recent study by Rice University sociologist Elaine Howard Ecklund found that 25 percent of scientists at elite universities—surely a population committed to most of the Relatives—believe in the existence of God. (For the population as a whole, the figure is something like 90 percent.) The novelist and spiritual thinker Marilynne Robinson is a strong believer in God, but her books, set in mid-twentieth-century Idaho, are filled with the moral complexities and uncertainties of the modern (and relative) world. The physicist Steven Weinberg is an atheist, but he believes in a "final theory" of nature, perfect and exempt from revision.

Similarly, these categories do not neatly separate religion from science. Although our religious traditions embrace most of the Absolutes, there are exceptions. For example, a fundamental belief in Buddhism is impermanence. The Reconstructionist branch of Judaism believes in constant change in the world and, furthermore, thinks of God not as an omnipotent Being but as the sum total of the natural processes that allow human beings to reach their highest aspirations.

Despite these exceptions, the Absolutes and the Relatives can be considered a large frame in which to view the dialogue between religion and science, or between spirituality and science. But I suggest that the issues go deeper, into the dualism and complexity of human existence. We are idealists and we are realists. We are dreamers and we are builders. We are experiencers and we are experimenters. We long for certainties, yet we ourselves are full of the ambiguities of the *Mona Lisa* and the *I Ching*. We ourselves are a part of the yin-yang of the world. Our yearning for absolutes and, at the same time, our commitment to the physical world reflect a necessary tension in how we relate to the cosmos and relate to ourselves. At the least, we are led to examine the differences and similarities between the physical world and what one might call the spiritual world. I have gone on that fraught journey myself. It is a winding and difficult path, with boundaries sometimes in clear view and sometimes dissolving into the mist. It is a journey sometimes of contradictions, and I may sometimes contradict even myself in these pages, depending on which forces are pressing me at the moment. I am a scientist, but I am not a swinging bob on a string.

Material

When I was seven or eight years old, growing up in landlocked Memphis, I visited my grandparents for a week at their little beach house in Miami. One dark and moonless night as I sat at the end of their dock, for some reason known only to children I grabbed a stick and stirred up the ocean beneath me. I was astonished to see the water shimmer with light. To my mind, the ocean was already a mysterious place, with its changing colors, its infinite gray skin stretching out to the sky, and its waves flowing in one after another, like the breathing of some large sleeping animal. But the glow of the seawater was magic of a different order. My imagination flared. Was this fairy dust? Was this some kind of galactic energy? What other secrets and powers lay below the ocean's surface? Excited, I ran into the house and commandeered my grandparents to witness the discovery. Again I stirred the water with my wand, and it happened again. Pure magic. I scooped up some of the supernatural liquid in a glass jar and took it into the house for further inspection. I'm not sure what I was hoping to find. What I did find, after the water settled, were tiny organisms floating about. In a dark room, they glowed faintly like fireflies. They felt slightly grainy in my hand. I was crestfallen. The magic was just little bugs in the water.

It is the end of June, and I am wandering about my small island in Maine. I've been thinking about the materiality of the world. Today, I just want to experience the fleshiness of this island. I run my hands along the bristly branches of a spruce tree. I could identify that prickle blindfolded. My bare feet sink into spongy moss. On the rocks lie mussel shells dropped from on high by crafty gulls seeking to break them apart and liberate the food within. The shells feel smooth and cool even in the sun. This tiny island in Casco Bay is shaped like a finger, half a mile long and a tenth mile wide. A high bony ridge runs down the spine of the island, a hundred feet above sea level, and my house lies on the north end of the ridge. To the south, there are five other cottages, cloaked from each other by a dense growth of trees—mostly spruce, but also pine, cedar, and poplar, whose leaves in the wind sound like hands clapping. The island's name is Lute—actually, that is my private name for the island, because of the natural music of the place. One will not find that name on any map of the region.

As did Thoreau in Concord, I've traveled far and wide on Lute Island. I know each cedar and poplar, each clump of beach rose, *Rosa rugosa,* each patch of blueberry bushes and raspberry brambles and woody stems of hydrangeas, all the soft mounds of moss, some of which I touch on my ramblings today. The tart scent of raspberries blends with the salty sea air. Early this morning, a fog enveloped the island so completely that I felt as if I were in a spaceship afloat in outer space—white space. But the surreal fog, made of minuscule water droplets too tiny to see, eventually evaporated and disappeared. It's all material, even the magical fog—like the bioluminescence I first saw as a child. It's all atoms and molecules.

The materiality of the world is a fact, but facts don't explain the experience. Shining sea water, fog, sunsets, stars. All material. So grand is the material that we find it hard to accept it as merely material, like meeting a man driving a Cadillac who claims he has only one dollar in his wallet. Surely, there must be more. "Nature," wrote Emily Dickinson, "is what we see / The Hill— the Afternoon / Squirrel—Eclipse—the Bumble bee / Nay—Nature is Heaven." In the last line, the poet leaps from the finite to infinity, to the realm of the Absolutes. It is almost as if Nature in her glory wants us to believe in a heaven, something divine and immaterial *beyond nature* itself. In other words, Nature tempts us to believe in the supernatural. But then again, Nature has also given us big brains, allowing us to build microscopes and telescopes and ultimately, for some of us, to conclude that it's all just atoms and molecules. It's a paradox.

For me, the human body is the most amazing and baffling phenomenon of the material world. How could it be that the exquisite and indescribable experience of consciousness, of thought and emotion, of the overpowering sense of an "I," is simply the result of so many electrical and chemical flows between neurons, which are themselves nothing but atoms and molecules? I am constantly struck dumb by this mystery. Surely, the first single-celled creatures moving about in the primeval seas did not have consciousness or thoughts. Evidently, those qualities emerged with increasing complexity and natural selection. As Darwin wrote in the last lines of his great book, "from so simple a beginning endless forms most beautiful and most wonderful have been, and are being evolved." Beautiful and wondrous, yes, but all material stuff, say the biologists.

Some years ago, I had a colonoscopy without being fully anesthetized and was able to watch on a computer screen the shifting views of the insides of my colon. I was dismayed. There, revealed in digital detail, was the deep interior of my body, a realm I had always considered as mysterious, a forbidden temple, fragile and secretive as it went about its profound business of keeping me alive. Surely that mystical place was separate from the world of tables and chairs, and gratefully hidden from the direct gaze of my eyes and my mind. But here it was, with no illusions. I was shocked to see ordinary flesh. I was shocked to see gelatinous membranes like quivering jelly, whitish in color, with bumps and ridges and turns. I felt like a trespasser in my own body. My colon was just material stuff. I am just material stuff. Or so I believe. I know this intellectually, yet I recoil from the idea.

Evidently, many others have also recoiled from the idea. A centuries-long debate in biology—and one not completely settled in some quarters even today— concerns whether living matter has some special quality not present in nonliving matter, some nonmaterial essence or spirit that is associated with life, especially intelligent life. The two sides of the debate have been called the "vitalists" and the "mechanists." Mechanists believe that a living creature is just so many microscopic pulleys and levers, chemicals and currents—all subject to the known laws of chemistry and physics and biology. Vitalists, on the other hand, argue that there is a special quality of life—some immaterial or spiritual or transcendent force—that enables a jumble of tissues and chemicals to vibrate with life. That transcendent force would be beyond physical explanation. Some call it the soul. The ancient Greeks called it *pneuma,* meaning "breath" or "wind." In the Christian Bible, *pneuma* means "spirit," as in "Verily, verily, I say unto thee, except a man be born of water and of the Spirit (*pneuma*), he cannot enter into the kingdom of God." *Qi* in the Chinese tradition. *Prana* in the Indian. In all of these cultures, this transcendent invisible energy, this *pneuma,* is associated with the magic of life. The *pneuma* rests cozily in the home of the Absolutes.

The debate between the vitalists and the mechanists is another form of the famous "mind-body problem": Is there an immaterial thing in living (and intelligent) creatures called the "mind" that is not reducible to the gooey tissues and nerves of the brain?

The ancient Greek and Roman "atomists" such as Democritus and Lucretius—proponents of the notion that everything was composed of atoms and only atoms—were of course mechanists. Spinoza was a mechanist. Plato and Aristotle were vitalists. They believed that an idealized "final cause," more spirit than matter, impelled a germ cell to develop toward an adult form. René Descartes, who famously articulated the separation between the intangible mind and the tangible body, proposed that in each human being an immaterial soul interacts with the material body in the pineal gland. Descartes was a vitalist. As was the great chemist Jöns Jacob Berzelius, author of *Lärbok i kemien,* the most authoritative chemistry textbook of the mid–nineteenth century. Berzelius expressed his views simply: "In living nature the elements seem to obey entirely different laws than they do in the dead."

For many vitalists, the *pneuma,* although invisible and immaterial, was responsible for supplying the energy needs of the body. And here, the vitalists found themselves unavoidably afoot in the land of the physical. Energy is physical. And since the mid–nineteenth century, the notion of energy has been owned by science. Science knows how to quantify the energy of motion, the energy of heat, the energy of gravity, the energy of molecules and chemical bonds, and the energy of everything else in the physical world. Science knows how to slice and dice energy into units called "joules" and "ergs" and "foot-pounds." Science knows how to tally up on a ruled accounting sheet the numbers for each swing of the arm, each inhale and exhale, each bead of sweat.

In the late nineteenth century, two German physiologists, Adolf Eugen Fick and Max Rubner, did just that for the human body. They tabulated the energies required for body heat, muscle contractions, digestion, and other physical activities and compared them against the chemical energy stored in food. Each gram of fat, carbohydrate, and protein is worth so many units of energy. After doing the arithmetic, the physiologists put away their sharp pencils and announced that the energy used by a living creature exactly equals the energy consumed in food—a victory not only for the mechanists but also for the law of the conservation of energy.

Still, many people remained dissatisfied. The thought that a human body might be reduced to so many coiled springs, balls in motion, weights and cantilevers, has not sat well with many of us. A dear friend of mine, a distinguished rabbi in Memphis, recently said to me, "I can't believe that we are just flesh. We have souls. But we're not bodies with souls. We're souls with bodies." Most religious leaders would agree. Beyond the religious domain, there's an entire field of alternative medicine called "energy medicine," which includes Reiki, qigong, and "biofield" medicine, all claiming to heal by working with hidden energies of the body.

Another point of view, eloquently stated in recent years by Marilynne Robinson, is affirmation of both body and soul but rejection of the dichotomy between the material body and the immaterial soul. (In fact, Robinson suggests even more broadly that the distinction between the physical world and the nonphysical world may be misguided.) Instead, she recommends that we acknowledge that the physical brain is "capable of such lofty and astonishing things that their expression has been given the names mind, and soul, and spirit."

The most extraordinary and graphic demonstration of the materiality of the body is the replacement of natural body parts by manufactures and machines. These days, we have artificial hands, artificial legs, artificial lungs, artificial kidneys, artificial hearts. In July 2001, the badly diseased heart of a telephone worker from Colorado named Robert Tools was cut out of his chest and replaced by the world's first self-contained artificial heart. Afterward, Tools lived for 151 days. The machine installed in his body was called an AbioCor artificial heart. It's about two pounds, the size of a cantaloupe. Made of translucent plastic and metal, the AbioCor looks like a tangle of auto engine cylinders fit together at odd angles. Blood is forced through the cylinders by a hydraulic pump and timed by an internal microprocessor. Wires extend down to the abdomen, where the little computer and lithium-ion battery are implanted. The internal battery can be charged remotely, so that no wires or tubes need penetrate the skin. After his initial recovery, Tools said of the thing in his chest: "It feels a little heavier than a heart . . . The biggest difference is getting used to not having a heartbeat . . . I have a whirring sound."

Scientists at the California Institute of Technology recently implanted two computer chips into the brain of Erik Sorto, age thirty-two, a quadriplegic. The output from the chips is connected to a computer, which interprets the patterns of their electrical activity, and the computer in turn is connected to a robot arm.

When Mr. Sorto is thirsty and *thinks* about reaching for a cup of water, the computer chips in his brain sense his desire, relay that thought to the computer, and the robot arm grabs a cup of water and brings it to his lips. It's all material.

I'm walking barefoot on my favorite part of Lute Island, the soft shoulder of a hill south of my house, partly shaded by spruce trees but with sun leaking through here and there. The ground is covered with a thick green mattress of moss, and when I lie down it yields to the shape of my body. I look up at the patches of blue and white between the trees, and I can hear sounds of seagulls and ospreys and the soft drone of a motorboat far, far away. If one listens, there's always music on this island. The waves rolling into the shore make cascades of sound, sometimes regular rhythms and sometimes duples and triples and offbeat syncopations—all set against the arpeggios and glissandos of the birds. Lying here in my moss bed and listening to the sounds, it's easy to drift off to sleep.

Although ours is a summer house, really not prepared for cold, I do occasionally venture here for short trips in winter. No roads or bridges connect Lute Island to the mainland, and my own boat is mothballed for the cold months, so I have to borrow a mainlander's skiff and navigate around the ice floes in Casco Bay, then pull up on the rocky shore. The island in winter is a German opera house in white, with white balconies and balustrades, white carpeted hallways, white winding stairways, white filigreed ceilings. The trees are expensive displays of Steuben glass, each branch lacquered with a transparent sleeve of crystal. When the wind blows through that forest of glass, you hear a high whining sound. If a fresh snow blankets the ground, it looks like cotton with a slightly bluish tint. If the snow has melted and refrozen and transformed to ice, each footstep sounds like breaking glass. The ice fractures and cracks. All of it is gorgeous, and all of it is ordinary material.

Few of my island neighbors come here in winter. The snow is untouched, my own footprints the only marks in the white except for the occasional track of a deer. Sometimes, I find the way to my favorite part of the island, the soft shoulder of hill south of my house, and I lie in the snow, looking up at the patches of blue and white between the ice-covered trees. If snow is falling, I can see the wondrous hexagonal pattern of each flake, a symmetry caused by the way that hydrogen and oxygen atoms bond together in their ballet of quantum physics. It's a dance of the atoms and molecules, choreographed by nature, and I sit in the plush seats of the German opera house and applaud. But there's too much to see. I rise from my seat and roam further out into the white. As I trek about the island in winter, I feel like an Arctic explorer.

A true Arctic explorer, Robert Edwin Peary, lived not far from here, on another small island, Eagle Island, about half the size of Lute. Some days in summer, I get in my boat and make a trip to Peary's island, which he bought in 1881. There, he built a magnificent house, which stands proudly on a stony prominence and looks out on the open ocean. Peary retired to Eagle Island in 1911, two years after his discovery of the North Pole. (Modern historians now believe that Peary did not in fact quite reach the North Pole, but was within sixty miles of it.) You can find Peary's crampons and snowsuits and other gear lying about his house, as well as photographs and letters. The house smells of old books and linseed oil. Peary penned his journal entries on blue-lined paper, in a handscript that slants from northwest to southeast. He too took note of the materiality of ice and snow. On April 6, 1909, the day that he claimed to have arrived at the pole, Peary wrote in his journal:

Weather thick . . . A dense lifeless pall of grey overhead, almost black at the horizon, + the ice ghastly chalk white with no relief. Like the ice cap, + just the thing an artist would paint for a Polar Icescape. Striking contrast to the glittering sunlit fields over which we have been travelling for 4 days, canopied with blue + lit by the sun + full moon. The going better than ever, hardly any snow on the hard granular surface last Summers surface of the old floes, the blue lakes larger. The rise in the temp. to −15° has reduced friction of the sledges −25% + gives the dogs appearance of having caught the spirits of the party.

Then:

The Pole at last!!! The prize of 3 countries, my dream + ambition for 23 years. Mine at last. I cannot bring myself to realize it. It all seems so simple + common place . . .

I try to imagine the "common place" experience of standing exactly at the pole of the earth (even if Peary was not quite there). I see myself perched on a glistening ball in space spinning about an imaginary axis through its center, and I am standing at the precise point where that axis emerges from the interior and punctures the ice. All other points on this ball, except at the opposite pole, are in motion. But I am still. You could say I am locally at rest. I am at rest relative to the center of the earth. But that center is itself in motion. As I stand here, that center hurtles around its central star at a speed of 65,000 miles per hour, and that central star, in turn, revolves around the center of the galaxy, the Milky Way, at a speed of 500,000 miles per hour. Do I know too much, or too little? I look up into space, as the cave dwellers did, and am transfixed by the infinite. Although I cannot touch it, I feel that I'm there. This resting yet unresting pole is quite a spot for viewing the universe.

Hummingbird

One summer, we had two or three hummingbirds darting about the feeder we hung on the veranda at the south end of our house. I quietly watched them from a back window, afraid of scaring them off. The birds seem to defy gravity. They hover. They float. They are "air within the air," said Pablo Neruda. They are a painter's accents, splashes of color on the canvas of the world, with iridescent blue heads and ruby red throats, iridescent light green and saffron bodies, orangish tails. You can't see the color of their wings because they are flapping back and forth 50 times per second. To supply oxygen for such an engine, the highest metabolism of almost all animals, the heart rate of a hummingbird is an incredible 1,300 beats per minute. Oxygen consumption per weight is ten times that of top human athletes.

You can actually calculate a lot of the design specs of the hummingbird from basic physics and biology. The hummingbird earns its living by being able to hover in midair while sucking the nectar from delicious flowers. How fast must it flap its wings to perform such an acrobatic feat? Slow-motion videos of hummingbirds show that their wings move in a rotary fashion, changing shape and angle throughout each cycle. If you require that the bird achieve the aerodynamic lift to support its weight, its wing tip must move at about 1,500 centimeters per second. That corresponds to a flapping rate of about 50 flaps per second, as observed.

You can also compute the required heart rate. A human being can flap its wings, i.e., arms, at about 4 times per second. (I have personally verified this result in front of my appalled students at MIT.) Since the heart rate of an exercising human is around 125 beats per minute, the required heart rate of an h-bird, needing 50 flaps per second, should be 50/4 times larger, or about 1,600 beats per minute. That number is pretty close to the figure observed. It's all a matter of science, like the swing of a pendulum. It's all material. But when I'm looking at the birds, suspended in space, I don't think numbers or gravity. I just watch and am amazed.

Stars

Does not the heaven arch itself there above?
—Lies not the earth firm here below?—
And do not eternal stars rise, kindly twinkling,
   on high?

FROM GOETHE, *Faust*

I have in my hand a little book titled *The Starry Messenger* (*Sidereus Nuncius* in its original Latin), written by the Italian mathematician and scientist Galileo Galilei in 1610. There were 550 books in the first printing of *Messenger*. One hundred and fifty still remain. A few years ago, Christie's valued each first edition at between $600,000 and $800,000. My paperback copy was printed in 1989 for about $12.

Although the history of science has not awarded *Messenger* the same laurels as Newton's *Principia* or Darwin's *On the Origin of Species,* I regard it as one of the most consequential volumes of science ever published. In this little book, Galileo reports what he saw after turning his new telescope toward the heavens: strong evidence that the heavenly bodies are made of ordinary material, like the winter ice at Lute Island. The result caused a revolution in thinking about the separation between heaven and earth, a mind-bending expansion of the territory of the material world, and a sharp challenge to the Absolutes. The materiality of the stars, combined with the law of the conservation of energy, decrees that the stars are doomed to extinction. The stars in the sky, the most striking icons of immortality and permanence, will one day expire and die.

Galileo was born in Pisa and grew up in Florence. From 1592, he taught mathematics at the University of Padua. Unable to discharge his financial responsibilities on his academic salary alone—he had to pay the dowries of his sisters in addition to supporting his three children by a mistress—he took in boarders and sold scientific instruments. In the late 1580s, he performed his famous experiments with motion and falling bodies. In 1609, at the age of forty-five, he heard about a new magnifying device just invented in the Netherlands. Without ever seeing that marvel, he quickly designed and built a telescope himself, several times more powerful than the Dutch model. He seems to have been the first human being to point such a thing at the night sky. (The telescopes in Holland were called "spyglasses," leading one to speculate on their uses.)

Galileo ground and polished his own lenses. His first instruments magnified objects a dozen or so times. He was eventually able to build telescopes that magnified a thousand times and made objects appear thirty times closer than they actually were. You can see Galileo's surviving telescopes in the rarely visited Museo Galileo, in Florence. His first one was 36.5 inches long and 1.5 inches wide, a tube made of wood and leather with a convex lens at one end and a concave eyepiece on the other. I recently looked out of a replica. First of all, I was surprised at how small the field of view was, appearing as a dime-sized circle of light at arm's length at the end of a long tube. And dim. However, after squinting for a while, I could indeed make out the faint images in that dime of dim light. And when I trained the primitive telescope on a building a hundred yards away, I could see details in the bricks not visible to my naked eye.

It's hard to imagine the thrill and surprise Galileo must have felt when he first looked up with his new instrument and gazed upon the "heavenly bodies"— described for centuries as the revolving spheres of the moon, sun, and planets. Beyond were the revolving crystalline spheres holding the stars, and finally the outermost sphere, the Primum Mobile, spun by the finger of God. All of it supposedly constructed out of aether, Aristotle's fifth element, unblemished and perfect in substance and form, what Milton described in *Paradise*

Lost (1667) as the "ethereal quintessence of Heaven."
And all of it at one with the divine sensorium of God.
What Galileo actually saw through his little tube were
craters on the moon and dark acne on the sun.

A few centuries earlier, Saint Thomas Aquinas
had successfully married Aristotelian cosmology with
Christian doctrine, including the ethereal nature of the
heavenly bodies and the notion that the earth stood
motionless at the center of the cosmos. (With one
Aristotelian idea Aquinas took exception: the lifetime
of the universe, infinite according to Aristotle, finite
according to Christianity.) Galileo's findings of imper-
fections in the heavenly bodies severely challenged the
Church. But the telescope itself was also a challenge.
Galileo's three-foot tube was one of the first instruments that amplified the human senses, that showed a
world not apparent to the natural eyes and ears. Nothing like this instrument had ever been seen before.
Many people were skeptical, questioning the legiti-
macy of the device and thus the validity of its findings.
Some regarded the strange tube as magical, not of this
world, as if a cell phone were presented to someone in
the year 1800. Galileo himself, although a scientist, did
not understand exactly how the thing worked.

We should recall that belief in magic, sorcery, and
witchcraft was widespread in Europe in the sixteenth
and seventeenth centuries. In just those two centuries,
40,000 suspected witches, most of them women, were

burned at the stake, hung from the gallows, or forced to put their heads on the chopping block. In 1597, King James VI of Scotland (who in 1603 became James I of England) complained about the "fearefull abounding at this time [and] in this Countrey, of these detestable slaves of the Divel, the Witches or enchaunters." It was believed that sorcerers could cast spells by damaging a strand of hair or a fingernail of an intended victim. Was the Italian mathematician's device a bit of sorcery?

Others regarded Galileo's telescopic findings with suspicion not because they reeked of black magic or contradicted theological doctrine but because they challenged personal worldviews and philosophical commitments. Cesare Cremonini, professor of Aristotelian philosophy at the University of Padua and a colleague of Galileo, denounced Galileo's claims of craters on the moon and spots on the sun but refused to look out of the tube. Cremonini was later quoted as saying, "I do not wish to approve of claims about which I do not have any knowledge, and about things which I have not seen . . . and then to observe through those glasses gives me a headache. Enough! I do not want to hear anything more about this." Another contemporary of Galileo, Giulio Libri, professor of Aristotelian philosophy at Pisa, also refused to peer through the tube. Galileo replied to these rejections in a letter to fellow scientist Johannes Kepler:

My dear Kepler, I wish that we might laugh at the remarkable stupidity of the common herd. What do you have to say about the principal philosophers of this academy who are filled with the stubbornness of an ass and do not want to look at either the planets, the moon or the telescope, even though I have freely and deliberately offered them the opportunity a thousand times? Truly, just as the ass stops its ears, so do these philosophers shut their eyes to the light of truth.

Galileo's little book is addressed to the Most Serene Cosimo II De' Medici, Fourth Grand Duke of Tuscany. The title page reads: "SIDEREAL MESSENGER, unfolding great and very wonderful sights and displaying to the gaze of everyone, but especially philosophers and astronomers, the things that were observed by GALILEO GALILEI, Florentine patrician and public mathematician of the University of Padua, with the help of the spyglass lately devised by him . . ." In his book, Galileo exhibits his own pen-and-ink drawings of the moon seen through his telescope, showing dark and light areas, valleys and hills, craters, ridges, mountains. He even estimates the height of the lunar mountains by the length of their shadows.

When he peered at the dividing line between light and dark on the moon, the so-called terminator, it was not a smooth curve as would be expected on the perfect sphere of theological belief, but a jagged and irregular line. "Anyone will then understand," Galileo writes, "with the certainty of the senses that the Moon is by no means endowed with a smooth and polished surface, but is rough and uneven and, just as the face of the Earth itself, crowded everywhere with vast prominences, deep chasms, and convolutions." Also reported were the sightings of moons around Jupiter, lending credibility to the notion that the other planets were similar to the earth. In other words, the earth was no longer special. All of which supported the proposal of Copernicus, sixty-seven years earlier, that the sun, rather than the earth, is the center of the planetary system. These were quite a few new ideas to pack into such a little book. And with no apologies to Aristotle or the Church.

Within a couple of months of the publication of *Sidereus Nuncius,* Galileo became famous throughout Europe—in part because the telescope had military and commercial value as well as scientific. (From "the highest bell towers of Venice," Galileo wrote to a friend, you can "observe sea sails and vessels so far away that, coming under full sail to port, 2 hours and more were required before they could be seen without my spyglass.") Word of the invention traveled by letter and mouth.

Galileo's announcement of dark spots on the sun was an even greater challenge to the divine perfection of the heavens. We now know that "sunspots" are caused by temporary concentrations of magnetic energy in the outer layers of the sun. Being temporary, sunspots come and go. In 1611, Christoph Scheiner, a leading Jesuit mathematician in Swabia (southwest Germany), procured one of the new gadgets himself and confirmed Galileo's sightings of moving dark spots in front of the sun. However, Scheiner began with the unquestioned Aristotelian premise that the sun was perfect and unblemished, and he went from there to proposing various precarious arguments as to why the phenomenon was caused by other planets or moons orbiting the sun rather than the sun itself.

As stated on the title page of his book, Galileo was a mathematician. Mathematics was generally regarded as existing in an abstract and logical world. Mathematics helped scholars calculate and predict the "real world," but it was distinct from that world. In particular, anti-theological models of the system of heavenly bodies were taken as merely calculational devices, describing *appearances* as opposed to *reality*. Thus, the earth-centered planetary system of Aristotle and Ptolemy and the sun-centered system of Copernicus could be placed on equal footing as calculational methods, as they both gave fairly accurate accounts of the positions of the planets. But the former accorded with theological and philosophical belief and was thus deemed to reflect reality.

When Galileo's observations became known, churchmen reacted with skepticism. On March 19, 1611, Cardinal Robert Bellarmine, head of the Collegio Romano, wrote to his fellow Jesuit mathematicians:

> I know that your Reverences have heard about the new astronomical observations by an eminent mathematician . . . This I wish to know because I hear different opinions, and you Reverend Fathers, being skilled in the mathematical sciences, can easily tell me if these new discoveries are well founded, or if they are apparent and not real.

Although the Church mathematicians argued about the details of Galileo's findings, they unanimously agreed that the sightings were real. Nevertheless, Galileo's telescopic findings and his support of the heliocentric model of Copernicus were considered an unpardonable attack on theological belief. For that offense, Galileo, a pious Roman Catholic who had once seriously considered the priesthood, was eventually tried by the Inquisition, forced to recant most of his astronomical claims, and spent the later years of his life under house arrest.

I want to focus now not on the displacement of earth as the center of the cosmos but on the newly conceived *materiality* of the heavens. Because it was that materiality, that humbling of the so-called heavenly bodies, that struck at the absolute nature of the stars. The demotion started with the observed craters and ruts on the moon. After 1610, dozens of thinkers and writers began to view the moon and planets as places of soil, air, and water, fit for human-like, if strange, habitation. In 1630, Johannes Kepler, the same fellow to whom Galileo wrote about the "stupidity of the common herd," finished work on a highly popular fantasy titled *Somnium* (Dream), in which a boy and his mother travel through space to the moon, called Levania. Everything in Levania is more extreme than on earth. On Levania, the mountains rise far higher than on earth, and the valleys plummet much lower. In the hot zone of Levania dwell living creatures, who are monstrously large and live only a single day. These animals, which swim, fly, and crawl, do not live long enough to build towns or governments, but they are able to find sustenance for life. Because Kepler was a distinguished scientist, his novel was taken seriously by the educated world and was read in the seventeenth, eighteenth, and even nineteenth centuries.

There were many other such fantasies. In "The Elephant Moon" (1670) by the poet Samuel Butler, self-satisfied gentlemen scientists, while observing the moon through a telescope, sight a battle of armies under way, during which a lunar elephant leaps from one line of soldiers to the other in mere seconds (possibly liberated by the reduced gravity of the moon). In 1698, the Dutch mathematician and scientist Christiaan Huygens wrote a book titled *The Celestial Worlds Discovered, or Conjectures Concerning the Inhabitants, Plants, and Productions of the Worlds in the Planets.* These books and poems were written for the general public. They give some sense of how people in the seventeenth century came to view the planets as ordinary material. Elephants do not rampage across divine spheres of ethereal quintessence.

But it was for the nature of stars that Galileo's findings had perhaps their most profound impact. The idea that stars might be suns had been proposed by the Italian philosopher and writer Giordano Bruno. In his *On the Infinite Universe and Worlds* (1584), Bruno wrote that "there can be an infinite number of other worlds [earths] with similar conditions, infinite suns or flames with similar nature . . ." (For his astronomical proposals as well as his denial of other Catholic beliefs, Bruno was burned at the stake in 1600.) By the early seventeenth century, various thinkers entertained the idea that stars might be suns. Thus, when Galileo reported blemishes on the sun, his findings had dramatic implications for all of the stars. The stars could no longer be considered perfect things, composed of some eternal and indestructible substance unlike anything on earth. The sun and the moon looked like other material stuff on earth. In the 1800s, astronomers began analyzing the chemical composition of stars by splitting their light into different wavelengths with prisms. Different colors could be associated with different chemical elements emitting the light. And stars were found to contain hydrogen and helium and oxygen and silicon and many of the other common terrestrial elements. Stars were simply material—atoms.

But "Nay," wrote Emily Dickinson.

Once Galileo and others had declared the stars to be mere material, their millennia were numbered—because all material things are subject to the law of the conservation of energy. This law is a paradigm of all laws of nature, both in its grand sweep of applicability and in its quantitative and logical formulation. Essentially, the law says that energy cannot be created or destroyed. Energy can change from one form to another, as when the chemical energy of a match turns into the heat and light of its flame. But the *total* energy in a closed and self-contained system remains constant.

Here's an example of how the law works. Suppose you have a sealed box with a fresh match in it, and let's say that the chemical energy in the match head is 3,200 joules. (The joule is a common unit of energy.) Now arrange for some device to strike the match. Some of the released energy goes into light, which bounces off the reflecting walls of the box until it is absorbed by air molecules in the box and raises their temperature. Some of the energy might go into heating water and making steam, which pushes on a piston and raises a paperweight a few inches. (Water, piston, and paperweight all reside within the closed box.) The match is dead. But its energy has not disappeared. It can be found elsewhere in the box. If you measure the increased heat energy of the air inside the box and the increased gravitational energy of the lifted weight, those increases will total 3,200 joules, exactly the energy in the unstruck match. You started with 3,200 joules of energy and ended with 3,200 joules of energy. That's the law of the conservation of energy.

One more important point about the law of the conservation of energy. Inherent in the law is the idea that all sources of energy are limited. If there were infinite sources of energy, the law would not work. No scale can weigh infinities; no calculator can tabulate infinities. It is quite possible that our physical world could not exist with infinite sources of energy.

Back to the stars. A star is like a giant match. It has a finite amount of energy stored within it—in the star, nuclear rather than chemical energy. That nuclear energy is released when atoms fuse together to make heavier atoms. But the supply of nuclear energy in a star is limited, just like the supply of chemical energy in a match. As the star "burns" its nuclear fuel, the energy is released into space, mostly in the form of light. If we imagine putting our star into a giant box, the total energy in that box remains constant, but the energy is gradually shifted from the star to the light in the box and the increased thermal and chemical energy of all things absorbing that light.

Of course, stars are not contained in giant boxes. But the principles remain. Stars, being physical material according to Bruno and Galileo and subsequent scientists, have a limited amount of energy. Stars radiate energy into space, thus depleting their finite supply of nuclear energy. Eventually that precious stellar commodity will be spent, at which point the stars will burn out and go dark. As will our sun, in about five billion years. In something like a thousand billion years, all of the stars in the sky will have gone cold. At that point, the night sky will be completely dark. And the day sky will also be completely dark. The myriad stars in the sky, once thought to be the final resting place of dead pharaohs, once thought to be the embodiment of constancy and immortality and other dispositions of the Absolutes, will eventually be cold floating embers in space.

Nature may at times appear to be a Painter or a Philosopher or a Celestial Spirit. But deep down she is a Scientist. She is quantitative. She is logical. And nothing better illustrates her ruthless and unyielding adherence to that logic than the law of the conservation of energy. Energy does not appear out of nothing. Energy does not disappear into nothing. The energy law is a sacred cow of physics. In all of the many different universes scientists have imagined, the total energy is constant in all of them. When in the early 1920s experiments revealed that the energies did not add up in the radioactive emissions of certain atoms, some physicists were so convinced of the inviolability of the law that they hypothesized the existence of new subatomic particles, invisible and undetected, which were furtively making off with the missing energy. A few years later, those subatomic particles, called neutrinos, were discovered. And the balance sheets were restored.

Two thousand years ago, the Roman poet and philosopher Lucretius suggested that the power of the gods over us mortals is limited by the constancy of atoms. Atoms could not be created or destroyed, said Lucretius. The gods could not make objects suddenly appear out of nothing or vanish into nothing because all things are made out of atoms, and the number of atoms remains constant. From Lucretius's epic poem "The Nature of Things" (*De Rerum Natura*):

> For surely a dread holds all mortals . . . because they behold many things happening in heaven and earth whose causes they can by no means see, and they think them to be done by divine power. From which reasons, when we shall perceive that nothing can be created from nothing, then we shall at once more correctly understand from that principle what we are seeking, both the source for which each thing can be made and the manner in which everything is done without the working of gods.

Lucretius's idea was a conservation law. The poet did not know how to tally up the number of atoms, as we tally up the number of joules in a box, but something was constant, and that constancy clearly provided great psychological comfort as well as understanding of nature. Let the gods and the supernatural have their sway, but they cannot alter the number of atoms here in our earthly world.

I would argue that the modern law of the conservation of energy also provides a kind of psychological comfort. With this law and others like it, nature can be made sense of. Nature can be calculated. Nature can be depended on. If you know the initial energy of the unstruck match and then measure the energy in the heated air, you *know* how high the weight must be lifted. The total energy is constant.

Ironically, we have traded one constancy for another. We have lost the constancy of the stars but gained the constancy of energy. The first is a physical object, the second a concept. Scientists cannot prove without a doubt that the total energy in a closed system is constant. But any violation of that principle would destroy the foundations of physics and suggest an unlawful universe. The idea of a lawful universe is itself an Absolute.

One last thought on this day of wanderings about my small island in Maine. The material of the doomed stars and the material of my doomed body are actually the same material. *Literally the same atoms.* Because all of the atoms heavier than the two lightest elements— hydrogen and helium—were manufactured in stars. When the universe was young, it was all hydrogen and helium. Then various clumps of gas gradually contracted into denser clumps, collapsed under their own weight, and formed stars. In the dense and hot nuclear furnaces at the centers of those stars, hydrogen and helium atoms fused together to form larger atoms: carbon and oxygen and silicon and beyond. Finally, some of those stars exploded and spewed their atoms into space. From where they coalesced to make planets.

From which single-celled organisms formed in the primeval seas. From which . . . It is astonishing but true that if I could attach a small tag to each of the atoms of my body and travel with them backward in time, I would find that those atoms originated in particular stars in the sky. Those exact atoms.

Atoms

It's a clear summer night and we've been sitting on our dock at Lute Island looking up at the stars. Overhead, the diaphanous white sash of the galaxy sweeps over the sky. And I feel myself falling into its depths. I am falling and falling and I am surrounded on all sides by the stars. I keep falling, further into space, until I am beyond the Milky Way. In the distance I see other galaxies, glowing spirals and pinwheels and elliptical blobs, each containing billions of stars. And I myself have grown larger. The galaxies have shrunk to mere dots. I see clusters of galaxies, then clusters of clusters, each appearing for moments and then dwindling away. I am a giant being striding through the dark halls of the cosmos, becoming larger and larger, but the universe is always still larger. Mansions within mansions. Space goes on and on and on, and never do I arrive at an edge. I am dizzy with infinity.

Then it reverses. I grow smaller. The clusters of gal-axies approach. Dots of light grow into galaxies. I see spirals and pinwheels and elliptical blobs of light. And I am still shrinking. Eventually, I find myself back in my home galaxy, the Milky Way. I can see individual stars, wispy nebulae. I continue to shrink and hurtle toward a particular star on the outskirts of my galaxy, then toward a particular planet, then toward the dappled brown coast of a landmass on that planet. Finally, I am sitting again on a wooden dock by the sea. But I continue to shrink. I go inside of a leaf, where I see green and blue vessels, veins and ridges, cellular lattices. Conglomerations of molecules. Then I see individual atoms, each a haze of electrical force. Atoms at last. The heralded atoms, the units of matter for centuries. Is this where my inward-bound journey will end? Have I arrived at the tiniest dots of reality? But there are smaller things still. I fall into a particular atom. I see quivering mists and vast empty spaces, then a dense throbbing mass down below at the core of the thing, the protons and neutrons, the nucleus of the atom. Relentlessly, I grow smaller. I enter a single proton. It is impossible, the violent energies nearly obstruct my view. Subatomic particles appear out of nothing like ghosts, then vanish. I see a trio of blurs, the three quarks. Have I finally reached the bottom of existence, the tiniest specks of the world? But there are smaller things still. I shrink within a single quark. I am blinded by energy. Far far off in the distance, many powers of ten smaller, I see vibrating strings of pure energy.

And, astoundingly, I continue to fall. I continue to fall. There's no end to it. I am dizzy with infinity, the infinity of the small.

It may have been the ancient Greeks who first conceived of a tiniest unit of matter, the atom or *atomos,* meaning "uncuttable." Atoms were not only uncuttable. They were indestructible. Atoms protected us from the whimsy of the gods, said Democritus and Lucretius, because atoms could not be created or destroyed. Even the gods had to obey atoms. Newton also prized atoms, but as the handiwork of God rather than as a defense against Him. Newton, who understood the logic of nature better than any mortal before him, wrote: "It seems probable to me that God in the beginning formed matter in solid, massy, hard, impenetrable, moveable particles . . . so hard as never to wear or break in pieces; no ordinary power being able to divide what God himself made *one* in the first creation." Indeed, atoms were the ultimate Oneness of the material world. Perfect in their indivisibility, perfect in their wholeness and indestructibility. Atoms were the embodiment of absolute truth. Atoms, along with stars, were the material icons of the Absolutes.

Atoms also unified the world. Because a leaf and a human being are made of the same atoms. Take apart a leaf or a human and we find identical atoms of hydro-

gen and oxygen and carbon and other elements. With
atoms, we have a foundation for material reality. On
that foundation, we can build *systems*. We can organize
and construct the rest of the world. Said Lucretius:
Pleasing substances are made of smooth and round
atoms, bitter substances of hooked and thorny atoms.
With atoms, we can make rules for the particular pro-
portions in which different substances combine, as
British chemist John Dalton did in the early nineteenth
century. Carbon monoxide: one atom of carbon joined
to one atom of oxygen. Carbon dioxide: one atom of
carbon joined to two atoms of oxygen. Never carbon
with one and a half atoms of oxygen. Because atoms
cannot be divided. With atoms, we can predict the
properties of the chemical elements, as Dmitri Men-
deleev did in the mid–nineteenth century.

Atoms prevent us from falling forever into smaller
and smaller rooms of reality. When we reach atoms—so
the thinking went—the falling stops. We are caught.
We are safe. And from there, we begin our journey
back up, building the rest of the world.

The idea of fundamental elements can be found in
all cultures and eras. Thinkers in ancient India conceived of a system of three "elements" for constructing
the cosmos: fire, water, and earth. Fire was associated
with bone and speech, water with blood and urine,
earth with flesh and mind. Aristotle built the cosmos
out of five elements: earth, air, water, fire, and aether

(for the heavenly bodies). For the ancient Chinese, the fundamental elements were wood, fire, metal, water, and earth.

Evidently, we humans are driven to construct the cosmos from basic elements. Why? Closely related: Why do we create systems and patterns? Are those patterns already there, independent of our desires, or do we impose them on a chaotic universe in order to scratch some existential itch? Could it be that we crave order for sanity? Another thought: With fundamental elements, we can conceive of the world as being *constructed*, whether the Constructor be an active God or the more passive Laws of Nature. A constructed world implies order and design. And the faint suggestion of an intelligence behind that order. Or perhaps that intelligence, in fact, is our own human intelligence, looking at ourselves simultaneously through both ends of the telescope.

At an engaging internet site hosted by the American Institute of Physics, you can listen to the voice of Joseph John Thomson talking about his discovery of electrons in 1897. Electrons were the first attack on the atom. At the time of the recording, in 1934, Thomson was seventy-eight years old and for many years the Cavendish Professor of Experimental Physics at Cambridge University. The recording crackles with static, but the words are unmistakable: "Could anything at first sight seem more impractical than a body which is so small that its mass is an insignificant fraction of the mass of an atom of hydrogen?" Impracticality indeed! But practicality is beside the point here. We're talking about a revolution of ideas, a bombing of the palace of Unity and Indivisibility. A photograph of Thomson at the time shows a deadly serious gentleman, balding, with spectacles and a thick walrus mustache, hands tightly clasped, starched white collar, staring unflinchingly into the camera as if he were looking two thousand years of history in the eye without apologies. "It was coming sooner or later," his gaze seems to say. "So buck up and take it like an adult."

Thomson made his discovery by measuring the paths of electrically charged particles as they were deflected by electric and magnetic forces. First, he and others had to develop good "vacuum pumps" for removing the air in the glass tubes through which the particles moved. Molecules of air interfere with the delicate trajectories of the tiny particles under study.

I have a great deal of respect for vacuum pumps. I used them myself during my short-lived encounter with experimental physics as a university student. A vacuum pump, when working properly, starts out with a coarse, grating sound, like the chug of a locomotive, then graduates to a clicking whine, rising in pitch, and ends with a smooth hum when a good vacuum has been attained. When the vacuum is incomplete, the pump never gets past the chugging locomotive stage.

The amount of deflection of a charged particle in a good vacuum indicates the ratio of its electrical charge to its mass. From previous experiments, Thomson and others already knew that particular ratio for hydrogen atoms, the lightest of all atoms. What Thomson found was that these other particles, the electrons—which he called "corpuscles" and which could be created by heating a piece of metal—had a ratio roughly 1,800 times larger than that of hydrogen atoms. Assuming the same electrical charge, the mass was then inferred to be 1,800 times smaller. Evidently, these things were really tiny compared to atoms. The atom was not the smallest unit of matter.

While Thomson was discovering the electron in England, Antoine Henri Becquerel and Marie Skło-dowska Curie were discovering the disintegration of atoms in France, what Curie called "radioactivity." Becquerel believed that the mysterious radiation recently observed to emanate from uranium, the so-called X-rays, were the result of the absorption of sunlight. The uranium X-rays, in turn, could be detected by nearby photographic plates. When Becquerel did his experiment, on February 26, 1896, Paris was cloudy. His uranium did not receive any energizing sunlight. On a whim, he decided to develop his photographic plates anyway. To his surprise, the photographic plates were strongly exposed, showing that the uranium emitted some kind of radiation on its own, without needing to be powered by the sun. Later experiments by Becquerel showed that the radiation consisted of electrically charged particles of some kind because they were deflected by magnetic fields, as were Thomson's electrons. After Becquerel's discoveries Curie did further studies of uranium rays and found that the uranium atoms were hurling out tiny pieces of themselves. A year later, Curie found the same atomic disintegration with another element, radium. The indivisible atom was, after all, divisible. And what lay inside? No one knew. The bottom of the universe had fallen out.

Here is the reaction of historian Henry Adams in 1903 to these disturbing developments:

> As history unveiled itself in the new order, man's mind behaved like a young pearl oyster, secreting its universe to suit its conditions until it had built up a shell of nacre that embodied its notions of the perfect. . . . He sacrificed millions of lives to acquire his unity, but he achieved it, and justly thought it a work of art.
>
> "One God, one Law, one Element" [Adams quoting Tennyson].

> Suddenly, in 1900, science raised its head and denied. . . . The man of science must have been sleepy indeed who did not jump from his chair like a scared dog when, in 1898, Mme. Curie threw on his desk the metaphysical bomb she called radium.

With his new corpuscles in hand, Professor Thomson proposed the "plum pudding" model of the atom: a tiny ball filled uniformly with a "pudding" of positive electrical charge, into which were sprinkled the negatively charged electrons. You needed the positively charged pudding to balance out the negatively charged electrons, since it was known that most atoms are electrically neutral.

Fifteen years later, the great physicist from New Zealand, Ernest Rutherford, and his assistants found that the atom was not a pudding at all. It was more like a peach. A hard nut resided at its center, containing all of the positive charge and nearly all of the mass. The new particles residing within that hard central nut were called protons and neutrons. Protons have positive electrical charge; neutrons have no charge. This peach picture emerged after Rutherford's team fired subatomic particles at a thin sheet of atoms. Some of the particles veered off at large angles, as if they had hit something hard, a hard nut in the atom. With pudding, the deflections should have been small. "It was quite the most incredible event that had ever happened to me in my life," boomed Rutherford. "It was almost as incredible as if you fired a 15-inch shell at a piece of tissue paper and it came back and hit you." The hard nut at the center of each atom, the "atomic nucleus," is a hundred thousand times smaller than the atom as a whole. To use an analogy, if an atom were the size of Fenway Park, the home stadium of the Boston Red Sox, its dense central nucleus would be the size of a mustard seed, with the electrons gracefully orbiting in the outer bleachers. In fact, 99.9999999999999 percent of the volume of an atom is empty space, except for the haze of nearly weightless electrons. Since we and everything else are made of atoms, we are mostly empty space. That vast emptiness is perhaps the most unsettling consequence of dividing the indivisible.

Eventually, Rutherford's protons and neutrons, at the center of the atom, would themselves be found to consist of even smaller particles called quarks.

Are we falling and falling without end? Are there unlimited infinities on all sides of us, both bigger and smaller? It is an unpleasant sensation. I am reminded of the Escher lithograph *Ascending and Descending*, which depicts a line of cloaked men walking around a quadrangle in a medieval castle. The disturbing feature of the picture, achieved through a trick of perspective, is that the walkers are always ascending, marching up a continuously rising staircase, and yet after completing the loop they end up exactly where they began. It is a staircase without beginning or end. It is a staircase that goes nowhere.

Escher made *Ascending and Descending* in 1960, at a time when physicists had recently discovered hundreds of novel subatomic particles in the new "atom smashers" and in high-energy radiation from space. The field of research into elementary particles and forces was thrown into chaos. In addition to the electrons and protons and neutrons, there were now delta particles and lambda particles, sigmas and xis, omegas, pions, kaons, and rhos, and more. When the Greek alphabet was exhausted for naming the new subatomic particles, the confounded physicists resorted to using Latin letters. Some of these particles had total lifetimes, from the moment they were created to the moment they disappeared, of a mere 10^{-21} seconds, or 0.000000000000000000001 seconds. Before, even with the sacred atom fractured, there had been some kind of order. There had been only the electrons and protons and neutrons. But now this howling zoo. There seemed to be no fundamental particles, no bottom to the infinite spiral down, no organizing principles.

Then quarks were discovered in the late 1960s. Temporarily, the plummeting stopped. Each of the hundreds of new particles could be understood as a particular combination of a half-dozen basic quarks. Quarks offered a new system for organizing the sub-

atomic zoo. Quarks were the new protons and neutrons, which, in turn, had been the new atoms. I
once asked physicist Jerry Friedman, co-discoverer of
quarks, whether he thought that the quark was the end
of the line, the smallest unit of matter. "Probably," he
answered. He gave reasons. But he hesitated. "I could
be surprised," he said with a grin. "There are always
surprises in science." Surprises in science are good
things, and bad.

The philosophers of ancient Greece developed a ter-
rifying view of the world called Zeno's Paradox. Suppose you want to walk 15 feet across a room. Before
you travel that distance of 15 feet, however, you must
go halfway, which is 7.5 feet. And before you go that
7.5 feet, you must travel half of that distance, 3.75 feet.
And before you go that 3.75 feet . . . And so on.

In their minds, the philosophers kept chopping
space into halves, into smaller and smaller dimensions
ad infinitum. The indivisible was pitted against the
divisible. The ultimate conclusion of this intellectual
exercise is that you cannot cross the room. In fact, you
cannot move even an inch. You are frozen in a meta-
physical conundrum. You are trapped by the infinity
of the small.

When scientists and mathematicians talk about
infinity, they are usually imagining a sequence of big-

ger and bigger spaces and numbers. But infinity can go in the other direction as well. Tolstoy commented on the problem in a nearly hidden appendix to his book *My Religion, On Life, Thoughts on God, On the Meaning of Life:* "The explanation of everything is sought in those [bodies] which are contained in the microscopic [bodies], and in those that are in them, and so forth, *ad infinitum* . . . The mystery will be revealed when the whole infinity of the small shall be fully investigated, that is, never."

Jerry Friedman, physicist rather than novelist, is more hopeful. He thinks that the quarks may be the end of the line. But there are things we don't know about quarks. Our current theory of quarks and electrons, called the Standard Model, is known to be incomplete. It does not include gravity. We do have a quite respectable theory of gravity, called General Relativity, but it has not been successfully wedded to the Standard Model. To perform that wedding, we must develop a theory of gravity that includes quantum physics, so-called *quantum gravity*. So far, all attempts have failed. But we have ideas about what such a theory would decree about the world within the quark, the infinity of the small.

General Relativity tells us that the geometries of space and time are affected by mass and energy. That is, a mass like the sun bends space the same way that a bowling ball on a trampoline sinks and flexes the rubber beneath it. Orbiting bodies, like planets, roll along that curved surface. A mass like the sun also makes time flow more slowly the closer you are to the mass.

Now, to the other partner in the marriage: quantum physics. Quantum physics shows that in the atomic and subatomic realm, particles take on a hazy, nondefinite character, behaving as if they existed in several places at once.

Although we don't yet have a theory of quantum gravity, we can still estimate the size of the region in which quantum physics and gravitational physics would both be important. Normally, quantum effects are significant only on very small scales, the size of atoms and smaller, and gravitational effects are important only on large scales, the size of planets and larger. But it turns out that on *really* small scales, both quantum and gravitational effects hold sway. This ultra-tiny scale is called the "Planck length," named after the physicist Max Planck, a pioneer in quantum physics. The Planck length is 10^{-33} centimeters, a hundred billion billion times smaller than a quark, which is itself a few hundred thousand times smaller than an atom. Another way to visualize the infinitesimal size we are talking about: the Planck length is smaller than an atom by about the same ratio as an atom is smaller than the sun. It is staggering what we can say about such infinitesimal elements of existence.

What are the implications of quantum gravity for

the infinity of the small? A critical aspect of quantum physics is that energy and other physical qualities do not come in continuous forms, like a flow of water from the tap, but in discrete units, like raindrops. The quantum raindrops are very, very tiny, so that we are not aware in our world that the flow is actually broken up into tiny parts. Some proposed theories of quantum gravity say that at the Planck scale, space is not continuous but exists in the form of indivisible cells, what one might call "Planck cells." Picture a tiny cube, one Planck length on a side. At the Planck scale, space is grainy. Space simply does not exist within a Planck cell. What we experience as space is the correlation between different corners of these cells.

Planck cells would be the atoms of space. Rather than atoms as the smallest unit of matter existing in space, we are now talking about the smallest units of space itself.

Surprisingly, some recent experiments have been able to probe nature at the ultra-tiny Planck length, or so the scientists claim. The graininess of space at the Planck scale should randomly slow down highly energetic light rays, affording a kind of friction as the light rays must jump from the corners of one Planck cell to the next. Less energetic light rays, with much longer wavelengths, would be oblivious to the Planck cells and move through space as if it were continuous, traveling at the ordinary speed of light, 186,000 miles per second. Researchers using NASA's Fermi Gamma-ray Space Telescope recently reported a *negative* result. That is, there was no slowdown of highly energetic light rays originating from a particular astrophysical explosion, compared to lesser-energy light rays coming from the same source. The scientists concluded that if space were indeed grainy, the cell size would have to be much smaller than Planck.

Regardless of whether space is indeed grainy at very small scales, physicists are confident that time and space must be chaotic at Planck. Because of the hazy, nondefinite character of quantum physics (called the Heisenberg Uncertainty Principle), at the dimensions of the Planck length space and time churn and seethe, with the distance between any two points wildly fluctuating from moment to moment, and time randomly speeding and slowing, perhaps even randomly going backward and forward. In such a situation, time and space no longer exist in a way that has meaning to us. The sensation of smooth time and space that we experience in our large world of houses and trees results only from averaging out this extreme lumpiness and chaos at the Planck length, in the same way that the graininess of a beach disappears when looked at from a thousand feet up.

Thus, if we relentlessly divide space into smaller and smaller pieces, as did Zeno, searching for the smallest element of reality, once we arrive at the phantasmagoric world of Planck, space no longer has meaning. At least, what we *understand* as "space" no longer has meaning. Instead of answering the question of what is the smallest unit of matter, we have invalidated the words used to ask the question. Perhaps that is the way of all ultimate reality, if such a thing exists. As we get closer, we lose the vocabulary. Sitting at midnight on my wooden dock by the sea and imagining myself falling and falling into smaller rooms of reality, I might continue to fall without limit, but once I reach Planck, space as I know it no longer exists. Space has been blown thin by an ancient glassblower, so thin that it dissolves into nothingness. The Planck world is a ghost world. It is a world without "time" and without "space." Perhaps that is where we should look for the Absolutes. But we no longer have words to describe the experience.

Ants

A psychiatrist friend, thinking I had nothing to do during my long summers in Maine, kindly gave me a thick book titled *Existential Psychotherapy,* by Irvin Yalom, emeritus professor of psychiatry at Stanford University. Professor Yalom is an acolyte of the Gestalt psychologists. Those thinkers hold that we naturally and unavoidably tend to organize all experience into meaningful patterns. When a picture of random dots is presented to us, we parse it into figures and background. When we see a broken circle, we mentally complete the circle. When we see odd behavior in people, we struggle to place it within some rational system. Professor Yalom writes that when incoming stimuli don't form patterns, "one feels tense, annoyed, and dissatisfied . . . We experience dysphoria [unease and anxiety] in the face of an indifferent, unpatterned world and search for patterns, explanations, and the meaning of existence." It is easy to see that the ability to find patterns could have been of survival benefit in our early days as *Homo sapiens*. But it could also be that the quest for patterns and systems is a part of the search for meaning, of trying to make sense of the world. Perhaps that deeper quest could also have motivated the idea of atoms and elements.

Over the years that I've been summering at Lute Island, I've come to realize that my main occupation here has not been reading or writing but figuring out whether *it* all adds up to anything. I will admit that the incoming stimuli are not forming patterns to my personal satisfaction.

A major obstacle is this (and now I am truly baring my material soul): I've always thought that for something to have *meaning*, it has to be permanent, or at least last a very long time. (I'm aware that a whole branch of philosophical thought deals with the question: What is the meaning of meaning?) Permanence is the Absolute that attracts me the most. What's the point, I ask myself, of anything that's here today and gone tomorrow—like a meal or a letter or a pair of shoes? By contrast, people still discuss and perform *King Lear* hundreds of years after it was written. People still gaze in awe at the ceiling of the Sistine Chapel. People still study the ideas of justice and government of Confucius and of Plato. Isn't that longevity a sure sign of meaning? I've always believed so. And, unconsciously, I measure my own strivings and the strivings of others on that basis. But I'm a materialist. And as a materialist I know that nothing lasts. Even *King Lear* might be forgotten in a thousand years. Or if a thousand isn't long enough for your personal idea of a long time, what about ten thousand years? Ten thousand years is the blink of an eye to the cosmos. Everything I see around me at this moment—the trees, my house, the books on my shelves, my children and their children and *their* children—will be gone without a trace in a few thousand years.

Sometimes I ask myself: Does meaning require some external agency, capable of recording events and precious moments in a permanent repository? God, if such a Being exists, could be that agency. Wouldn't any other agency also pass away after a certain lapse of time? What if we had a second external agency, grander and far longer-lived than the first, and suppose all the information and meaning recorded by the first agency was eventually inherited by the second? Yet this new arrangement would save the situation for only a limited time. Because the second agency, being finite, would also pass away after a time.

In fact, does *anything* we do on our modest planet— only one among billions of planets in our galaxy, which is only one among billions of galaxies in the observable universe—have any meaning on a grand scale? What do creatures on planet XUFK, a thousand galaxies away, know or care about *anything* that happens on earth? Unless there exists an infinite and permanent observer such as God—some absolute authority or scaffold by which to judge and preserve meaning—then the situation seems hopeless to me. On the other hand, perhaps my starting assumption, that meaning requires permanence, is erroneous. Or perhaps meaning itself is an illusion. After all, why should I insist on meaning? Fish and squirrels get by quite well without it.

These are the kinds of unsettling thoughts that run through my mind while I'm walking about Lute Island or sitting at my writing desk. I should probably be doing something more profitable, like collecting clamshells or scratching out bad sentences. But I can't help myself. My uneasy mind is both blessing and curse. Sometimes I pose to myself the following situation, which I'll call the Smart Ant Conundrum: Imagine a colony of highly intelligent ants. Suppose further that this ant colony lasts for a hundred years. Normal ant colonies last only twenty years or so, when the queen trundles off to spawn another colony, but let's assume that a long dynasty of queens have followed each other to replenish this particular colony. Each individual ant lives only a year, so there have been many generations of ants in this colony. This is an old colony. Over the century, these brainy ants create a great civilization. They build advanced structures underground. They compose music. They create paintings and theater. They write books and record histories of their society. They develop science and make theories about the cosmos, both inside the anthill and beyond. They have emotions and intimate relationships. Then one day, a flood comes and totally destroys the ant colony. Totally. There is nothing left—no ants, no ant books, no ant paintings, no remnants. Nothing. Everything is completely destroyed. There's no trace left in the universe of this magnificent ant colony. The question I ask myself: Did the ant colony have any meaning? And now, after the colony is gone, with no record of its existence, does it have meaning?

I squint my eyes and try to see Professor Yalom's patterns. I think I see them in the small, when I watch reflected sunlight flickering on my bedroom ceiling or hummingbirds hovering near my front porch. I think I see patterns and meaning in the moment. But not beyond. Maybe the moment is all there is. Maybe I should just gather my clamshells and be quiet. The exquisite experience of joy—when I am completely consumed by a pleasurable activity such as conversation with good friends or good food or laughing with my children—is certainly one of the moment. But for some reason, I and many of my fellow travelers are not satisfied with the moment. The Now isn't enough. We want to go beyond the moment. We want to build systems and patterns and memories that connect moment to moment to eternity. We long to be part of the Infinite.

Monk

In recent years, I've gotten to know a prominent Buddhist monk in Cambodia by the name of Yos Hut Khemacaro. His friends call him Khema. He was born in 1948 in a little farming village in the province of Prey Veng and went to a primary school there administered by monks. At the age of ten, as he now vividly recalls, he was "attracted to learn wisdom" and began studying Buddhism. Eventually he was ordained a monk himself. In 1973, Khema started working with the United Nations on human rights, in Australia and Thailand. After the devastation of the Khmer Rouge genocide in the mid- to late 1970s, during which monks were targeted along with all educated people, Khema returned to Cambodia and played a major role in rebuilding the Buddhist monkhood there.

I visited Khema one warm day in January at Wat Lanka, his monastery on a busy avenue in Phnom Penh. I was hoping that he might help me fathom my communion with the stars that summer night in Maine and other experiences I'd not understood. Buddhism embodies an interesting mix of beliefs. The Four Noble Truths would appear to reside within the realm of the Absolutes, while the Buddhist doctrine of impermanence is a Relative.

Wat Lanka is a large temple complex containing several pagodas, patios and walkways, and living quarters for some two hundred monks. The magnificent front gate rises forty feet high and is guarded by stone lions on both sides. As soon as you step through that arched edifice, you leave behind the steady drone of motors and the shouting of street sellers—and enter a realm of serenity. Slowly, I walked past the gold-leaf pagodas. I passed obelisk-like stone stupas and scattered stone pots filled with red and pink bougainvillea. I passed through courtyards with young men in orange robes quietly strolling in pairs. Eventually, I came to Khema's living quarters, a tiny house at the back end of the complex. We sat under some trees. A faint scent of jasmine wafted through the air.

Under the trees, Khema and I began discussing modern physics and cosmology. I had brought him one of my own books on the subject. "Buddhism is in complete agreement with science," Khema said slowly and smiled. Then he added, "Science puts in more details." Khema explained the Buddhist belief that the universe has gone through an infinite number of cycles in the past and will go through an infinite number of cycles in the future. When I mentioned to him that modern cosmologists have evidence that the universe will continue expanding without further cycles, he laughed. Perhaps he thought it was preposterous that science could know such a thing. Or perhaps he thought it delightful that science could know such a thing. While we were talking, Khema's sister, an ancient nun with a shaved head, appeared from somewhere and silently served us tea. I noticed that her hands were wrinkled and worn, like the cracked yellow paint on the walls of Khema's house.

I asked Khema how Buddhists know that the universe has already gone through an infinite number of cycles. He said that knowledge comes from the Buddha, one of whose names is *lokavidū*, the "knower of worlds." "The Buddha knew everything," said Khema. He took out a pen and scribbled down some books I should read. His writing was slow and deliberate, like himself. We stopped talking. In the distance, I could hear the rise and fall of monks' chanting, soft like the sound of the wind, unintelligible. I had no idea what time it was.

Truth

During my visit with Khema, he mentioned the Four
Noble Truths of Buddhism: First, that life is filled with
suffering. Second, that the origin of suffering is the craving and clinging to impermanent things. Third, that
the suffering of life can be ended. And fourth, that the
path to that end is through meditation, self-discipline,
and mindful living. Although the Buddha first articulated the Four Noble Truths 2,500 years ago, Khema
was careful to make clear that we come to these truths
through our own experience with the world. But on
other matters, such as their belief in the infinite cycles
of the universe, Buddhists base their convictions exclusively on the words of the Buddha, a human being born
as Siddhārtha Gautama, later known as the *lokavidū*,
the "knower of worlds." I thought to myself: How do
we know that the Buddha was the knower of worlds?
Were Einstein and Darwin also knowers of worlds?
The truths and laws that we believe about the physical

and spiritual worlds—why do we believe them? And on what authority?

The concept of a law goes back at least four thousand years. Long before laws for the physical world, the ancient Assyrians articulated their Code of Ur-Nammu. Those first laws were, of course, rules for behavior in human society. Quantifiable only in the number of shekels of silver owed or quarts of salt poured into the mouth for each specified infraction. For example: "If a man proceeded by force and deflowered the virgin slavewoman of another man, that man must pay five shekels of silver."

The Four Noble Truths of Buddhism—are they laws? Perhaps they are simply observations of the human condition. Certainly religious traditions have rules governing behavior, similar to the Code of Ur-Nammu. Not that human beings will *always neces-sarily* behave according to certain rules, as a dropped stone will necessarily fall to the ground. But various theological traditions command us human beings to behave according to certain rules. For example, "Thou shalt not kill," the sixth of the Ten Commandments. Or, from the Qur'an: "He [Allah] loves those who keep themselves pure and clean . . . When ye prepare for prayer, wash your faces, and your hands (and arms) to the elbows; rub your heads (with water); and (wash) your feet to the ankles." Such daily routines as the manner of washing before prayer may seem mundane and insignificant, but when they are described in the Qur'an and considered the word of Allah, they are elevated to laws. Likewise, the statement that the relief of mortal suffering is to be had through meditation might seem like an opinion or a bit of philosophy or a page from a self-help book. But when it is uttered by the "knower of worlds," it takes on the imperative of a law, or an absolute truth. (Here and in the following chapters, I use the words "law" and "truth" interchangeably, with the recognition that I am not being quite precise. I take a law to be a statement that expresses a truth. In science, laws are almost always expressed in quantitative and mathematical form.)

Science and religion differ profoundly in the way that truths are discovered. In religion and theology, these truths and beliefs seem to have two origins. First are the sacred books, such as the Bible, the Qur'an, the Vedas, the Pali Canon, and their interpretations. Believers assume that these books contain the true word of God or of special enlightened beings. If so, the authority of the teachings derives from the infinite wisdom associated with God or the Buddha or other divinities. That divine authority can also be transferred to the authority of the religious institution as a whole, as in the authority of the "Church" in Catholicism, or the authority of shariah in Islam. The second origin of truth is more personal, what one might call the "transcendent experience," which I will discuss more in the next chapter.

Quotations from the sacred books are used to declare truths ranging from the origin of the universe to the question of free will to the details of reproductive biology. For example, Saint Thomas Aquinas makes a difficult philosophical argument against the Aristotelian view that the universe has existed forever, but then falls back on Scripture for his authority:

> Potentiality is prior in time to actuality (although actuality is prior in nature), yet, absolutely speaking, actuality must be prior to potentiality, as is clear from this, that potentiality is not reduced to actuality except by some actual being. But matter is being in potentiality. Therefore God, first and pure actuality, must be absolutely prior to matter, and consequently cause thereof. This truth divine Scripture confirms, saying: *In the beginning God created heaven and earth.*

Another leading Christian theologian, John Calvin, invokes Scripture to argue that everything that happens in the (physical) world, including the actions of human beings, is predetermined by God.

As we know that it was chiefly for the sake of mankind that the world was made, we must look to this as the end which God has in view in the government of it. The prophet Jeremiah exclaims, "O Lord, I know that the way of man is not in himself: it is not in man that walks to direct his steps," (Jeremiah 10:23) . . . man can do nothing without the power of God . . . Scriptures moreover, the better to show that every thing done in the world is according to His decree, declare that the things which seem most fortuitous are subject to Him.

According to the Islamic hadith, the Prophet Muhammad taught these facts of reproductive biology:

He is created of both, the semen of the man and the semen of the woman. The man's semen is thick and forms the bones and the tendons. The woman's semen is fine and forms the flesh and blood.

Still today, many religious thinkers attribute absolute authority and absolute truth to the teachings of the sacred books, called "divine revelation." Here is part of the announcement of the Second Vatican Council's Dogmatic Constitution on Divine Revelation, called the *Dei Verbum* (in English: "word of God") and endorsed by Pope Paul VI:

> The books of Scripture must be acknowledged as
> teaching solidly, faithfully and without error that
> truth which God wanted put into sacred writings
> for the sake of salvation.

I respect the notions of God and other divine beings.
However, I insist on one thing. I insist that any statements made by such beings and their prophets about
the *material world,* including statements recorded in
the sacred books, must be subject to the experimental
testing of science. In my view, the truths of such statements cannot be assumed. They must be tested and
revised or rejected as needed. The spiritual world, and
the world of the Absolutes, have their own domain.
The physical world should be the province of science.

Transcendence

For me, as both a scientist and a humanist, the transcendent experience is the most powerful evidence we have for a spiritual world. By this I mean the immediate and vital personal experience of being connected to something larger than ourselves, to feeling some unseen order or truth in the world. The experience I had looking up at the stars off the coast of Maine was a transcendent experience. I've had others. The transcendent experience is beautifully described by a clergyman in William James's book *Varieties of Religious Experience* (1902):

> I remember the night, and almost the very spot on the hilltop, where my soul opened out, as it were, into the Infinite, and there was a rushing together of two worlds, the inner and the outer. It was deep calling unto deep—the deep that my own struggle had opened up within being answered by the unfathomable deep without, reaching beyond

the stars. I stood alone with Him who had made me, and all the beauty of the world, and love, and sorrow, and even temptation. I did not seek Him, but felt the perfect union of my spirit with His . . . Since that time no discussion that I have heard of the proofs of God's existence has been able to shake my faith. Having once felt the presence of God's spirit, I have never lost it again for long. My most assuring evidence of his existence is deeply rooted in that hour of vision in the memory of that supreme experience.

The feelings described here are echoed in a letter that the French novelist and dramatist Romain Rolland wrote to Sigmund Freud in 1927 about something the writer called the "oceanic feeling" (*sentiment comme océanique*). A decade earlier, Rolland had won the 1915 Nobel Prize in Literature for, according to the citation, "the lofty idealism of his literary production and to the sympathy and love of truth with which he has described different types of human beings." In the letter to Freud, Rolland suggested that the source of religious energy lies in an "oceanic feeling," which is a "sensation of eternity, a feeling of something limitless, unbounded—as it were oceanic . . . a feeling of an indissoluble bond, of being at one with the external world as a whole."

The transcendent experience may or may not involve a supreme Being, or God. And here I would like to remind the reader that there are many different notions of God within the various philosophical and theological traditions. Some, like the transcendentalists and pantheists, consider God to be synonymous with nature and all that exists. Others, like the classical deists, consider God to be an all-powerful and purposeful Being, but a Being that does not intervene in the physical world once created. Reconstructionist Jews view God as the sum total of all natural processes that inspire human beings to be the best they can be. The most common view of God is that of an all-powerful and purposeful Being that does, from time to time, intervene in the physical world.

Regardless of which belief you subscribe to, the transcendent experience, unlike the received wisdom acquired from the sacred books, is intensely personal. And the authority of that experience and the understandings gained from it rest in the experience itself. No other person can deny the validity of what you have felt. The feelings cannot be disproved.

The transcendent experience as an avenue to truth is a deeply human path. In fact, some philosophers and theologians believe that truth exists *only* in the realm of humanity and the human mind. Such a belief runs head-on against the scientific belief that truth also exists independent of our minds. Doesn't the parabolic arc of a tossed stone exist whether or not a human mind conceives such a curve? These opposing views were highlighted in an extraordinary conversation between Einstein and the great Bengali artist and philosopher Rabindranath Tagore. The physicist and the poet admired each other and arranged to meet on July 14, 1930, at Einstein's house in Caputh, near Berlin. Here is a portion of their conversation:

E. Truth, then, or Beauty is not independent of Man?

T. No.

E. I agree with regard to this conception of Beauty, but not with regard to Truth.

T. Why not? Truth is realized through man.

E. I cannot prove that scientific truth must be conceived as a truth that is valid independent of humanity; but I believe it firmly. I believe, for instance, that the Pythagorean theorem in geometry states something that is approximately true, independent of the existence of man. Anyway, there is a reality independent of man, there is also a truth relative to this reality . . .

T. Truth, which is one with the Universal Being, must essentially be human; otherwise whatever individuals realize as true can never be called truth, at least the truth which is described as scientific and which only can be reached through the process of logic, in other words by an organ of thoughts which is human.

E. The problem begins whether Truth is independent of our consciousness.

T. What we call truth lies in the rational harmony between subjective and objective aspects of reality . . .

Laws

According to the ancient Roman historian Plutarch, the Greek mathematician and scientist Archimedes was so entranced with a particular geometrical diagram that he didn't notice when the attacking Romans penetrated his city. The year was 212 BC. The Second Punic War was on fire, and the Roman consul Marcus Claudius Marcellus had been raining stones and missiles on the walled city of Syracuse from his sixty-galley fleet in the harbor.

Eventually, the city fell. Plutarch describes Archimedes' love of mathematics and science as "a divine possession," to the extent that the mathematician frequently forgot to eat or bathe until forced to by friends. As the siege of Syracuse was nearing its end, an invading Roman soldier came upon Archimedes quietly contemplating a mathematical problem and commanded him to go to a meeting with the victorious consul. Archimedes replied that he'd rather not at the moment—until he'd finished his calculations. At this point the soldier, enraged, ran the mathematician through with his sword.

Little is known of Archimedes' life. But we do know that he was one of the first human beings to formulate a law of the physical world, his "law of floating bodies" (ca. 250 BC):

Any solid lighter [less dense] than a fluid will, if placed in the fluid, be so far immersed that the weight of the solid will be equal to the weight of the fluid displaced.

We can speculate on how Archimedes arrived at his law. At the time, balance scales were available for weighing goods in the market. The scientist could have first weighed an object, then placed it in a rectangular container of water and measured the rise in height of the water. The area of the container multiplied by the height of the rise would give the volume of water displaced. Finally, that volume of water could be placed in another container and weighed. Undoubtedly, Archimedes performed this exercise many times with different objects before devising the law. He probably also performed the experiment with other liquids, like mercury, to discover the generality of the law. Although Archimedes invented many practical devices, such as systems of pulleys to lift great weights and machines of war, I cannot think of much commercial or military utility in his law of floating bodies. The scientist seems to have been drawn to it solely by his fascination with the physical world and for the pleasure of his mind.

All laws of the physical world are like Archimedes' law. They are precise. They are quantitative. And they are general, applying to a large range of phenomena. Perhaps it is astonishing that nature should obey laws at all. On the other hand, it is quite possible that the physical universe could not exist without laws, that an unlawful universe would involve a fatal self-contradiction or logical inconsistency, like $2 + 2 = 4$ *and* $2 + 2 = 3$. Certainly, an unlawful universe would be a frightening place in which to live. Wheelbarrows might suddenly float in the air or planets change orbits without warning.

In the last two hundred years, we have discovered laws that govern the behavior of electricity and magnetism, the forces inside atoms, the expansion of the universe, and many other phenomena. From those laws, we have been able to explain in quantitative detail everything from the color of the sky to the orbits of planets to the weight of a flying bustard to the six-sided symmetry of snowflakes. And we have seen no evidence to contradict the notion that *all* phenomena in the physical world are governed by laws.

Here's a wonderful law of nature that you can verify for yourself: Drop a weight to the floor from a height of 4 feet and time the duration of its fall. You should get about 0.5 seconds. From a height of 8 feet, you should get about 0.7 seconds. From a height of 16 feet, about 1 second. Repeat from several more heights, both bigger and smaller, and you will discover the rule that the time exactly doubles with every quadrupling of the height, a rule found by Galileo in 1590. Galileo's "law of falling bodies" can be expressed mathematically as $t = 0.25 \times \sqrt{d}$, where t is the falling time in seconds and d is the distance fallen in feet. The number 0.25 comes from the gravity of the earth and derives from its size and mass. With this rule, you can now predict the time to fall from any height. You have discovered firsthand the lawfulness of the physical world.

The law of the swinging pendulum that I confirmed at the age of twelve was a version of Galileo's law of falling bodies. It wasn't true because I had read it in *Popular Science,* or because it was something I wanted to believe, or because the famous Galileo had announced it nearly four centuries earlier. The law was true because it worked. Evidently, it described a fundamental property of the physical world.

Let me sketch the methods and manner of science. Galileo discovered his law of falling bodies by letting objects slide down an inclined plane and timing their fall with a water clock. (Why the inclined plane? It had the benefit of slowing the fall and making measurements easier.) Galileo's law is actually a statement that the acceleration of any falling body near the earth's surface is constant. The law was later found to be a special case of Isaac Newton's more general laws of motion and gravity, published in 1686. His gravitational law says that the gravitational force between two masses is proportional to the product of the masses and inversely proportional to the square of the distance between them. Newton discovered *his* law by analyzing the orbits of planets in the gravity of the sun, orbits that had been carefully charted by a previous astronomer.

For two centuries, Newton's law worked beautifully. But in the mid–nineteenth century, with the increased precision of telescopes and careful measurements, astronomers concluded that the orbit of the planet Mercury didn't quite match the predictions of the law. The accumulating discrepancy was extraordinarily tiny, about one-hundredth of one angular degree *every century*. In almost all disciplines of human interest, such a minuscule disagreement would be ignored, like being a penny off in a bank balance of a hundred thousand dollars. However, Newton's law was so precise and well defined, and the measurements of Mercury so precise, that some scientists were troubled. Then, in 1915, Albert Einstein proposed a new theory of gravity called General Relativity, elegant in its geometrical construction. (For those few readers who would like to see what General Relativity looks like in mathematical form, here it is: $R_{\mu\nu} - \frac{1}{2}Rg_{\mu\nu} + \Lambda g_{\mu\nu} = 8\pi T_{\mu\nu}$.) Einstein's theory completely explained the orbit of Mercury. Furthermore, it predicted many new phenomena, such as the deflection of starlight by the sun's gravity, black holes, and gravitational waves—the latter were first detected in 2015 with the LIGO experiment. Now the main point: Despite its subtlety and enormous success, we know that Einstein's theory will also need revision, as I discussed in a previous chapter.

One can view the project of science as a progression of discovering ever more accurate descriptions of nature. Those descriptions, which we express as "laws of nature," are mathematical tools that allow us to make predictions—how far the needle on a voltmeter will move when hooked up to a particular electrical circuit, or how long it will take for a group of uranium atoms to disintegrate. The proposed laws are judged on the accuracy of their predictions. We scientists do not pretend to know what "reality" is. That slippery idea is either unknowable or a resident of the land of philosophy. The project of science is to make accurate predictions about the measurements of rulers and clocks, the outcome of experiments.

All the laws of nature discovered by scientists are considered provisional. They are considered to be approximations of deeper laws. The laws are constantly being revised as new experimental evidence is found or new (and testable) ideas are proposed. In fact, what we call "laws of nature" should really be called "approximate laws of nature." But that expression is a mouthful.

As mentioned in the last chapter, there are major differences in the truths of science and religion and the manner in which those truths are discovered. Unlike religion, science does not accept truths and laws based on the authority of divine beings or their designated emissaries, or even from the institution of science as a whole. The ideas of great figures in science, such as Niels Bohr and Alexander von Humboldt and Claude Bernard, might be taken seriously for a time simply out of deference to those towering intellects, but eventually the ideas will be accepted or rejected based on experimental test. Similarly, the personal transcendent experience, while a vital source of truth in religion, is viewed with suspicion in science. Personal passion may be a motivation and pleasure for scientists in doing their work, but the only truths and results that are accepted by the scientific community are those that can be reproduced in different experiments by different scientists and rederived by different people from the same mathematical equations. Indeed, science goes out of its way to try to eliminate the personal from the process of acquiring knowledge.

Finally, we see strong differences between science and religion in the process of *revision*. The core beliefs of religion—and indeed the ideals of any of the Absolutes—are not subject to change or revision. The wisdom of God or of the enlightened Buddha is absolute. It is perfect. So is the nature of permanence, of unity, of indivisibility, of absolute truth. These entities of the Absolute are not approximations, like Newton's equations for gravity. They are exact. They are like perfect circles, eternal and unassailable. They are like the *amrita*, the elixir of immortality. They are like Plato's ideal forms.

Doctrine

I had the good fortune of being a graduate student in physics in the early 1970s, when the first black hole was discovered. It was called Cygnus X-1, and it was located about 7,000 light years from earth. In other words, it would take a light beam 7,000 years to get from there to here. At that time, a small number of students were beginning their doctoral thesis research on the astrophysics of Cygnus X-1 and other black holes. We worked with Einstein's equations for gravity and other equations describing gas flow, radiation processes, and thermodynamics. Never once did we question the validity of those equations in describing a bizarre phenomenon 7,000 light years away. We also applied those equations to black holes in other galaxies, hundreds of millions of light years away, and to events that happened in the infant universe, 14 billion years ago. Again, we didn't question the validity of the equations for phenomena far far away from our comfortable desks and teakettles on earth.

Without ever hearing it spoken out loud, we budding scientists simply embraced a principle I call the Central Doctrine of Science: All properties and events in the physical universe are governed by laws, and those laws hold true at every time and place in the universe. Graduate students in science absorb this belief through every pore of their skin. It is an unconscious but powerful commitment. The reliable applicability of laws at every time and place is an important part of the Doctrine. In the physical world, the laws of nature cannot apply to some phenomena but not to others, or apply at some times but not at other times. It is not OK with me if the principles of aerodynamics work on some of my airplane flights but suddenly quit on others.

There are a couple of implicit assumptions within the Doctrine. The laws governing the physical world should have a mathematical form. From our many centuries of studying and quantifying the workings of nature, we have learned that the language of nature is mathematics. We assume that to be a truth about the physical world. Second, the laws cannot raise more questions than they answer. For example, a law for gravity that has a different mathematical form for each planet in the universe would not be acceptable to scientists, since it would leave unexplained why every planet was subject to a different kind of gravitational force. The Central Doctrine of Science assumes that the laws governing the physical world have a certain generality and completeness.

The Central Doctrine, as I have defined it, raises the subtle but important question of what we mean by "the physical universe." (In these pages, I use "physical universe" and "material universe" interchangeably.) If we say that the "physical universe" refers to all matter and energy, as commonly assumed, we have simply transferred the question to a definition of matter and energy. Physicists agree on a definition of matter and energy, but that definition might not satisfy everyone. Some people might regard the "soul" or the "spirit" as energy, but it would not be an energy necessarily measurable by thermometers or radio antennae. So I must resort to a circular definition. The physical universe is all material and phenomena for which the Central Doctrine of Science applies. There are other realms of existence to which the Central Doctrine does not apply—realms that we might call the "nonphysical universe" or the "spiritual universe" or the "ethereal universe." In this sense, the reach of science is limited.

That said, the concept of a physical universe, obeying logic and laws, has been extremely successful in the last five hundred years. Vaccines and antibiotics, radio and television, computers and iPhones are just some examples. One example from physics represents the dramatic success in our ability to understand and quantify the physical universe: Detailed mathematical calculations using the latest theories in quantum physics *predict* that the magnetic strength of the electron, a type of subatomic particle, is 1.159652182, while the value measured with highly sensitive equipment is 1.159652181. Surely, there is something material out there, beyond our minds, that we understand very, very well. Because such an enormous range of phenomena—from the orbits of planets to the color of the sky to the structure and manipulation of DNA—have yielded to the methods of science, it would be a mistake to take the Central Doctrine merely as a tautology. The physical world, even if limited, is vast. One might restate the Central Doctrine in the following way: A huge range of phenomena are lawful and subject to the analysis of science.

I call the Central Doctrine of Science a doctrine because, despite its success, it cannot be proved. It must be accepted as a matter of faith. No matter how lawful and logical the material cosmos has been up to now, we cannot be sure that something illogical, unexplainable, and fundamentally unlawful might happen tomorrow. Scientists accept the Central Doctrine on faith. Our faith in the Central Doctrine is so strong that when we find physical phenomena that cannot be explained in terms of current laws, we attempt to revise those laws rather than abandon our belief in a lawful universe.

When it was found that the orbit of Mercury could not be completely explained in terms of Newton's law of gravity, scientists did not attribute the discrepancy to an unsolvable mystery or to the breakdown of order in the physical world or to the intervention of a whimsical god. Instead, they recognized a physical problem that required a more advanced physical understanding. Similarly, when it was found in the early twentieth century that the law of the conservation of energy seemed to be violated in the emissions of certain atoms, in a process called "beta decay," scientists did not abandon their faith in the Doctrine. Instead, they proposed that some previously unknown and invisible particles were carrying away the missing energy. Eventually, those predicted particles, called "neutrinos," were found. In fact, I cannot imagine *any* event in the material world that would cause most scientists to label the event a miracle, unexplainable by science. If a wheelbarrow began to float, a scientist would look for magnetic levitators or, if necessary, assign the phenomenon to some new kind of force—a natural and lawful force, not a supernatural force.

In sum, I would argue that despite the big differences between religion and science in the way that knowledge is obtained and revised, both share a degree of faith, a belief and commitment to the unprovable. Science experimentally tests and verifies all of its beliefs about particular phenomena. But it cannot do so with the fundamental belief, the Central Doctrine. The Central Doctrine must simply be accepted. In that sense, the Central Doctrine of Science is one of the Absolutes.

There is one more Absolute in science: the "final laws," or the "final theory." Many, if not most, physicists believe in a "final theory" of nature, a theory beyond approximation. Such a belief has not emerged from science itself. On the contrary, as we have discussed, the history of science portrays a long progression of continuing revision, in which new theories replace older ones, with the upstarts retained for a while until they are themselves replaced by even more accurate theories. Despite this history, many physicists believe in a "final" theory. Such a final set of the laws of nature would need no further revision. It would be perfect. *But we would never be able to prove it was final,* because we could never be certain that a new experiment or phenomenon the next day might clash with the theory and require its further revision. In other words, even if we were in possession of a final theory, we would never know it. Yet we believe.

In his book *Dreams of a Final Theory* (1992), Nobel Prize–winning physicist Steven Weinberg writes: "Our present theories are of only limited validity, still tentative and incomplete. But behind them now and then we catch glimpses of a final theory, one that would be of unlimited validity and entirely satisfying in its completeness and consistency . . . My own guess is that there is a final theory, and we are capable of discovering it."

A final theory is really an extension of the Central Doctrine of Science, that nature is completely lawful. As Weinberg implies, our approximate theories are becoming finer, grander, and more stunning in their mathematical beauty. Let me give just one example, involving the work of the great English theoretical physicist Paul Dirac. Dirac was a man of few words. In fact, his colleagues at Cambridge University defined a unit of conversation called the "dirac," which was one word per hour. In the late 1920s, Dirac wrote down on a piece of paper an elegant equation for the behavior of electrons, an equation that combined Einstein's relativity (not Einstein's General Theory of Relativity, a theory of gravity, but his Special Theory of Relativity, a theory of motion and time) with the newly discovered quantum physics. Given the requirement of mathematical and logical consistency, there were very few ways to formulate such an equation; it was almost like filling in the blanks of a crossword puzzle. Unexpectedly, Dirac's equation predicted the existence of new subatomic particles, now called positrons, identical to electrons but with opposite electrical charge. A few years later, positrons were discovered. In an article in *Scientific American*, Dirac wrote:

> It seems to be one of the fundamental features of nature that fundamental physical laws are described in terms of a mathematical theory of great beauty and power, needing quite a high standard of mathematics for one to understand it. You may wonder: Why is nature constructed along these lines? One can only answer that our present knowledge seems to show that nature is so constructed. We simply have to accept it. One could perhaps describe the situation by saying that God is a mathematician of a very high order . . .

Theoretical physics is a temple built of mathematics and logic and aesthetics. Working and living in that temple is a mystical experience of its own. It is not surprising to me that many of my colleagues believe in a final theory of nature, a theory of absolute perfection. Perhaps that perfect theory is the ultimate reality of the philosophers. Or perhaps it's the physicists' version of Nirvana.

Motion

One of my favorite paintings is van Gogh's *Starry Night*. Bordering on the psychedelic, it depicts the night sky just before dawn above a little village in Provence. Dim houses sleep peacefully in the village, nestled near a church with a turquoise steeple. In the distance, we see deep purple hills sloping down to the town; in the foreground, a dark cypress tree, its branches sweeping upward like a black flame. But the center stage of this painting is the sky. Buttery stars melt out of the night, each with exaggerated haloes in white, blue, and green. The moon is a bold orange crescent framed against a yellow disk. And two strange swirls roll through the sky like galactic waves. In fact, the entire sky appears as a cosmic whirlpool. Space, given solid form with van Gogh's thick brush strokes, flows and bends around each star as if caught in its dominion of gravity and light. The sky churns in flamboyant grandeur, while the little community of human beings sleeps quietly below. *Starry Night* is not a painting you look at unstirred.

Van Gogh's painting means many things to me, both for its suggestions of mortality in the face of the infinite and also for the tortured life of the man who created it. The artist made the painting in mid-June 1889—looking east from his second-floor bedroom window in the Saint-Paul-de-Mausole lunatic asylum. Just a month earlier, he had checked into the asylum voluntarily after a mental breakdown in which he famously cut off part of his ear. Astronomers have verified that the moon on that date could not possibly have looked as it did in van Gogh's painting. Furthermore, there was no village in that location near the asylum. However, a particularly large and bright "star" in the painting has been identified with Venus, which was indeed visible at that time and in that place. All in all, the painting is a mixture of fiction and fact, like most artistic creations. I believe that in his happier periods van Gogh must have been in a constant state of rapture, akin to the moments I've experienced looking up at the stars on dark nights in Maine. Although van Gogh ultimately rejected organized religion, he once wrote to his brother Theo that he had a "tremendous need for, shall I say the word— for religion—so I go outside at night to paint the stars." A year after creating *Starry Night,* van Gogh committed suicide. He was thirty-seven.

There's more to the story. In 1990, medical research-ers published an article in the *Journal of the American Medical Association* diagnosing at least part of van Gogh's mental condition as arising from Ménière's disease, also called vertigo, a disorder of the inner ear. Vertigo causes you to feel that you or the world around you is moving, when actually everything is at rest. Sometimes you feel that you're falling, or spinning. You get dizzy.

The concepts of motion and stillness are more com-plex than they seem. The metaphorical usage is clear. Shakespeare's Othello says to Montano: "The gravity and stillness of your youth the world hath noted," sug-gesting Montano's groundedness and dependability. A motionless person, not flitting about here and there, can be taken in, reckoned and reasoned with, relied upon. In one of her poems, Emily Dickinson compares stillness to a "smooth mind." Stillness connotes peace-fulness, centeredness, quiet, equilibrium, balance—the certainty that one has fully comprehended the situa-tion and arrived at a harmony with the world. When we join God, we arrive at the final stillness, eternal rest. Stillness and complete rest are notions of the Absolutes.

Then there's the physical. In Aristotle's cosmology, everything had its proper place and proper motion. The proper place of the earth was at the center of the universe. And the earth itself was thought to be absolutely at rest, against which all other motions could be gauged. The heavenly bodies, including the other planets and stars, revolved about the unmoving earth. Then came Copernicus and Galileo and the sun-centered planetary system. For Newton, who accepted the proposition that the earth was in motion, there remained something absolutely at rest, namely the body of God. Yet Newton never discussed how one might scientifically determine movement relative to God. His own physics forbade any notion of absolute rest. In his equations for mechanics, for example, only the *relative* motion between two objects could be measured or have physical meaning. Newton, both devout believer and master logician of the physical world, simply stated his theological views without proof.

As it turns out, the contrasting notions of motion and stillness unavoidably propel us into the nexus between science, religion, and the nature of the physical world.

It is a warm day in July, and I am lying on the mossy shoulder of a hill on Lute Island looking up. At this latitude, I figure I'm spinning at a speed of about 750 miles per hour as the planet whirls on its axis. But I don't feel it. There's no wind rushing past my ears. That's because the earth's air is being dragged along with the rotating planet. And the acceleration is tiny. So, it's only my mind that knows I'm in motion. In a similar manner, it must have been an intellectual shock for the spectators who gathered in the Meridian Room of the Paris Observatory one day in 1851 to observe the lazy rotation of the plane of a giant pendulum. According to the laws of physics, that rotation was proof that the earth spun on its axis. A journalist who attended the performance wrote in *Le National* newspaper the next day: "At the appointed hour, I was there, in the Meridian Room, and I saw the Earth turn." But what did the journalist actually see, or feel? Nothing more than I do, lying on the mossy shoulder of a hilltop in Maine. That journalist was *told* that the rotation of the plane of the pendulum was an illusion. In actuality, the plane of the pendulum was fixed. The rotating thing, the journalist was told, was the table beneath the pendulum. It was the earth that spun.

And what is the earth spinning relative to? Relative to the stars, said Heraclides of Pontus and Nicholas Copernicus and Isaac Newton. Are the stars fixed in space? Or do *they also move*? And move relative to what? Consider this: I'm lying on my back on the mossy slope of a hill on Lute Island, a tiny bauble of land off the northeast coast of America. I am at rest relative to the ground. But the ground is spinning about at 750 miles per hour relative to the center of the earth. And the earth is orbiting the sun at a speed of about 65,000 miles per hour. And the sun is orbiting the center of the Milky Way galaxy at a speed of about 500,000 miles per hour. And the Milky Way is moving through inter-galactic space at a speed . . . Trying to figure it all out, one motion on top of another, makes your head spin.

Could it be that there is no such thing as absolute rest? Absolute rest might also be a creation of the human imagination, an expression of our yearning for Shakespeare's groundedness, or for Emily Dickinson's smoothness of mind. Personally, I am content knowing that at this moment, lying on my back on a mossy hill in Maine, I am at rest relative to the ground.

In the nineteenth century, the notion of absolute rest reappeared—not in the form of a unique astronomical object but as a ubiquitous fluid called the ether, filling up all space. It was the always mischievous theoretical physicists, working with pencil and paper, who pro-posed the existence of this nearly weightless and invis-ible stuff, spreading out like a gossamer ocean through every cubic inch of space. If this cosmic ocean existed, one could by experiment measure any motion relative to it, just as one can measure the speed of a boat mov-ing through water. The ether, since it filled up all space, could be deemed completely motionless, without ref-erence to anything else. In other words, the ether con-stituted absolute rest. Each planet or star or anything else could then be said to move at so many miles per hour through the ether. And some objects might be at rest in the ether, like a boat on its anchor. But where did this ether come from, and why?

Just at the end of the American Civil War, a Scottish theoretical physicist named James Clerk Maxwell unified the theories of electricity and magnetism with a set of four beautiful equations beloved ever since by every student of physics. I learned Maxwell's equations in my sophomore year of college. Among many other things, Maxwell's equations show that the phenomenon we experience as light is in fact a wave of oscillating electrical and magnetic forces traveling through space. That's where the ether comes in.

All waves known in the nineteenth century required a material medium through which to propagate and move. For example, sound waves need air or some other material medium. Take the air out of a room, and you can scream all you want but no one will hear you. In fact, a sound wave in air is nothing more than air molecules being pushed toward each other, then pulled apart, then pushed together in a Slinky-like pattern that travels from one place to another. The same with water waves. Water waves are moving ripples of water. Take away the water and there can be no water waves. By analogy, it was thought that light would require a material medium, the ether, to transport it from one place to another. Furthermore, since we see light from distant stars, the ether would need to occupy all space.

Various experiments attempted to measure the

speed of the earth through the ether. As the earth orbits the sun, so the reasoning went, it moves through the ether just as a boat moves through a lake. Light waves sent out in different directions from the orbiting earth should travel at different speeds. Only a boat at rest in the water sees no variation in the speed of waves traveling outward from it in different directions. To their consternation, the experimenters found that the speed of the light waves was always the same, no matter their direction.

In 1905, a young patent clerk named Albert Einstein, twenty-six years old at the time, suggested that perhaps the ether didn't exist—in which case the notion of absolute rest was a dream. You couldn't say how fast a planet or anything else was moving in any absolute sense because there was no fixed and ubiquitous frame of reference. Only the *relative* motion of two objects, like a planet and its central star, was measurable. Einstein's ideas went much further than that. In following the logical trail of his proposal, the audacious young patent clerk realized that *time,* too, must be only relative. According to Einstein, the belief that time is absolute—that a second is a second is a second always and everywhere—is based on a false impression of the world. Einstein was like Neo, the main character of the film *The Matrix,* who slowly realizes that something is terribly wrong with his perception of reality. Neo's brain, and the brains of all humans in that world, had been commandeered by a master computer. Our brains have not been taken over by a computer, but our clumsy sensory perception has led us into illusions about the behavior of time and space. Einstein claimed that even if all observers have identical clocks, the elapsed time between two events depends on the motion of the observers relative to those events. In other terms, clocks in motion relative to each other do not tick at the same rate. The greater the relative speed, the greater the discrepancy. Why haven't we human beings noticed such bizarre behavior? Because the discrepancies are tiny unless the relative motions are huge, far beyond what we experience in daily life.

I've sometimes tried to imagine the young Einstein and his thoughts as he made his momentous discoveries about space, time, and motion. Where did he get the courage and fearlessness, the self-confidence, even the insolence, to challenge the understanding of *time*? Where in history could one find a comparison? Alexander the Great? Copernicus? Martin Luther? Marcel Duchamp? I imagine the cramped patent office in Bern, Switzerland, lodged within the four-story stone building on Speichergasse. In my mind, I can see a half-dozen wood desks and tables in the room, sagging bookshelves with folders of patent descriptions, a scuffed wooden floor, inkwells, pens, a clock on the

wall. By 1900, Einstein had finished his undergraduate
degree but had failed to be accepted into any graduate program in physics. Evidently, he had not received
favorable recommendations from his college professors, to whom he refused to bow and scrape as other
students did. And he did not always hide his opinions
of his esteemed professors. In 1901, Einstein sub-
mitted his doctoral thesis (which he had conceived,
researched, and written completely on his own) to
Professor Alfred Kleiner at the University of Zurich. In
that thesis, Einstein criticized the work of a colleague
of Kleiner's. A month later, in a letter to his sweetheart,
Mileva Marić, Einstein wrote: "Since that bore Kleiner
hasn't answered yet, I am going to drop in on him on
Thursday . . . This sort [Kleiner] instinctively considers every intelligent young person as a danger to his
frail dignity . . . But if he has the gall to reject my doctoral thesis, then I'll publish his rejection in cold print
together with the thesis & he will have made a fool of
himself."

From a young age, Einstein had been a rebel and a
loner. His sister Maja remembered that until he was
seven or eight years old, whenever Albert was asked
a question he would mutter the answer twice, once
silently to himself and then aloud, as if needing to hear
it first in his own inner world. At the age of sixteen,
the boy renounced his German citizenship. He hated
the harsh discipline of his teachers at the German gym-

nasiums, dropped out of high school, and moved to Milan with his family. The young Einstein felt further embattled when his parents rejected Mileva (whom he later married). Also a physicist, Mileva had a congenital limp and was three years older than Albert. "My parents are very distressed about my love for you," Einstein wrote to her in 1900. (He was twenty-one years old at the time.) "Mama cries bitterly, and I am not given a single undisturbed moment here. My parents mourn for me almost as if I had died." Unable to pursue graduate studies in academia, Albert continued reading and thinking physics on his own. He was living close to poverty, barely supporting himself with meager tutoring jobs. Finally, in early 1902, he was offered a job at the Swiss patent office in Bern. Here, he could have a regular income. Here, he did not have to suffer his bloated professors. Here, he would have the solitude to think about physics and nature, between quick and brilliant analyses of the patent applications that passed across his desk.

To make a revolution in thought may often require a young person, one not yet in the vise of inherited knowledge and views of the world. Darwin was twenty-eight when he first conceived the theory of natural selection. Picasso was twenty-six when he invented cubism. A rebellious nature might also be helpful. Einstein had few loyalties and commitments—none to pre-existing science, none to social norms of behavior, none to his teachers, none to organized religion, none to his country. None to any of the Absolutes (except to his faith in a lawful cosmos). Einstein rejected the possibility of life after death, immortality, permanence in the material world. If he believed in God, it was not a personal God, concerned with the affairs of human beings. Yet Einstein, like Spinoza, did believe in a majestic order in the universe. One evening in the mid-1920s, he described those beliefs at a dinner party in Berlin (recorded in the diary of another guest): "Try and penetrate with our limited means the secrets of nature and you will find that, behind all the discernible laws and connections, there remains something subtle, intangible, and inexplicable. Veneration for this force beyond anything we can comprehend is my religion." A few years later, Einstein gave an even more explicit statement of his spiritual views in an interview with George Sylvester Viereck, a poet and German propagandist. The interview took place in Einstein's Berlin apartment, where Elsa, a cousin whom he married after divorcing Mileva, served raspberry juice and fruit salad. Viereck asked Einstein point-blank if he believed in God. Einstein's reply:

I'm not an atheist. The problem involved is too vast for our limited minds. We are in the position of a little child entering a huge library filled with books in many languages. The child knows someone must have written those books. It does not know how. It does not understand the languages in which they are written. The child dimly suspects a mysterious order in the arrangements of the books but doesn't know what it is. That, it seems to me, is the attitude of even the most intelligent human being toward God. We see the universe marvelously arranged and obeying certain laws but only dimly understand these laws.

Einstein had discovered the relative world. He had abolished absolute motion, absolute time, and absolute space. Yet he did believe in one Absolute. He believed in a beautiful and mysterious order underlying the world, a "subtle, intangible, and inexplicable" force. Many scientists, whether they believe in a personal God, an impersonal God, or no God at all, have experienced the "mysterious order" that Einstein describes, often in transcendent moments that have some similarity to the transcendent experiences I've discussed previously. Unfortunately, few scientists have commented on these experiences. Werner Heisenberg, who at age twenty-five discovered the foundations of quantum physics, described in his autobiography the moment when he realized that his new theory would work. Recovering from an illness, he had taken a two-week leave of absence from the University of Göttingen, traveled alone to a German archipelago on the North

Sea, and was doing nothing but taking daily walks, going on long swims, and thinking science. "At first, I was deeply alarmed. I had the feeling that, through the surface of atomic phenomena, I was looking at a strangely beautiful interior, and felt almost giddy at the thought that I now had to probe this wealth of mathematical structures that nature had so generously spread out before me. I was far too excited to sleep." James Watson described the moments when he realized that the structure of DNA must be a double helix: "As the clock went past midnight I was becoming more and more pleased. There had been far too many days when Francis [Crick] and I worried that the DNA structure might turn out to be superficially very dull, suggesting nothing about either its replication or its function in controlling cell biochemistry. But now, to my delight and amazement, the answer was turning out to be profoundly interesting."

Surely these experiences are partly the exhilaration of solving a hard problem after months or years of struggle, or the delight of the artist in creating something new. Yet they feel like more. As I lie on this mossy hill on an island in Maine, hurtling through space in my mind, I return to the rooms of my own scientific discoveries, certainly modest compared to those of Einstein or Heisenberg or Watson. But real. A memory: For several months, I had been working in seeming futility on a problem dealing with the creation of particles

and their antiparticles in ultra-high-temperature gases.
Matter from energy. I was stuck. What time of year
was it? I think it was June. The house was hot. I remember it was a Sunday morning. I woke up about five a.m.
and couldn't sleep. Something was happening in my
mind. I was thinking about my physics problem, and
I was seeing into it. My head was lifting off my shoul-
ders. I felt weightless. I was floating. And I had no sense
of my self, where I was, or who I was. I did have a feeling of rightness. I had a strong sensation of seeing into
this problem and understanding it and knowing that I
was right. In the next room, my two-year-old daughter
squirmed and coughed in her bed; my wife was still
sleeping. While I was skimming across the water like
a smooth stone. Somehow, I was also sitting quietly at
the kitchen table with the pages of my calculations,
wrinkled and coffee-stained, daylight just beginning to
come through the window like the end of my dream.
I was alone, and that's what I wanted. I needed no
help. I was alone with the problem, seeing into it. And
I could suddenly see how different pieces fit together,
in a fashion surprising and familiar at the same time—
with a certain *inevitability*, like a Shakespearean sonnet
in which no word can be changed. I was swollen with
energy. And I saw it in the equations—the curve of the
thing, the mathematical solution. But this was not only
mathematics. Somewhere in the universe, in another
galaxy perhaps, this thing might be happening, accord-

ing to these equations. How could that be? What had I touched?

Such moments are common in all creative activities, in the arts and in the sciences. But in the sciences, there is a vital connection to the physical world. The transcendent moment in science is both inner and outer. You travel deep into your own imagination and being, and you are totally alone with the experience. At the same time, you find something that is larger than yourself, that exists in the world outside of yourself. It is a double discovery. It is a discovery of something inside of your mind and also of something out there in the world, a pattern, a law, a piece of the fabric of nature. Despite the relativity of the world, you sense something enduring, something at rest. You connect—but only partially. You feel Einstein's mystery. Where did the books in the huge library come from?

Centeredness

For thousands of years, the notion of absolute rest was part of a worldview in which the earth resided at the center of the cosmos. Aristotle argued that there couldn't be two such centers, because then earth and all earth-like materials would be confused about which center to occupy, contradicting the observations that tossed stones clearly know where to fall. If earth indeed has some special position, it naturally follows that all things were constructed for us—if not for our positive benefit then at least with us earthlings as a prominent part of some cosmic design. Centers do not happen by accident. A cosmic plan with us as the center then leads to the concept of a personal God, a God who cares about human beings as individuals.

Most people in the world today, in fact, do believe in a *personal* God. A 2008 survey by the Pew Research Center found that 60 percent of American adults believe that "God is a person with whom people can have a relationship." A similar survey by the National Opinion

Research Center found that 67 percent of Americans agreed with the statement that "there is a God who concerns Himself with every human being personally." In his book *Warranted Christian Belief,* the prominent Christian thinker and philosopher Alvin Plantinga writes that classical Christian belief holds that "God is a person: that is, a being with intellect and will . . . a person [with] affections, loves, and hates."

Recent scientific discoveries seem to challenge the idea of a personal God. Certainly they challenge the idea that the universe was made for us human beings. Data from the Kepler astronomical satellite, launched in 2009 and specifically constructed to search for planets in the "habitable zone"—that is, the right distance from their central star to possess liquid water—suggest that something like 10 percent of all stars have at least one "habitable" planet. At the beginning of 2017, another astronomical satellite, the NASA Spitzer Space Telescope, found seven habitable planets around a single star, only forty light years from earth (close by in galactic terms).

There are several hundred *billion* stars in our galaxy alone, and a hundred billion galaxies just within the observable universe. Overwhelmingly, the odds favor life forms elsewhere in the universe. Although we do not know in detail how life developed on earth, the odds that no life exists on the billions and billions of other habitable planets would be as improbable as no fires *ever* starting in a billion trillion dry forests. Almost certainly life elsewhere in the universe would not be like ours. But biologists and even perhaps artists and philosophers would recognize it as life. And with so many life-bearing worlds and billions of years of cosmic evolution, there must be a range of civilizations, some less advanced than ours and some more.

And now back to the question of a personal God. There's no plausible reason why our particular civilization on planet earth should be more or less worthy of attention than the billion trillion others, each possibly with its own version of the Moses story and the Abraham story and the Krishna story. Not to mention the billions of creatures living in each of the billion trillion civilizations. A personal God would be extraordinarily busy with such a huge congregation and so many souls to attend. But then again, we have no conception of the capacities of God.

Death

Early August. I am lying in a hammock on the veranda of our house on Lute Island. (Just now, it occurs to me that I spend an inordinate amount of time on this island sitting or lying about, without accomplishing much of anything, although this morning I did uproot an annoying tree stump by patiently digging up the soil underneath each of its thick roots and then severing the roots one by one with my chain saw until the once fearsome stump was just a naked and helpless thing, robbed of its strength, and I picked it up like it was a baby's toy and tossed it into the woods.) But back to the hammock. I have come to believe that it's good to waste time. In fact, it's probably essential to waste time. That's when the mind has a chance to think about what it wants to think about, without being cudgeled and shoved by the external world.

My wife, a painter, and I have been coming to this small island for twenty-five years to waste time. Nothing is absolutely motionless, says Einstein, but I'm centered in this island. Wherever it goes, hurtling through space as the earth orbits the sun and the sun orbits the galaxy, I go with it. I've planted myself here, like the *Rosa rugosa* down the hill, stubborn and thorny. At this moment, I can hear the call of a gull and the wind blowing through trees like the sound of a distant waterfall and the tiny purr of a boat engine far off in the bay. Then there's the steady and slight sound of the waves, playing counterpoint to the soft music of birds. But all of it slips into the silky silence of this place. I embrace that silence. I breed silence and am bred by it. On this island, I am light years away from the noise and heave of the world. Like Thoreau, I came here "because I wished to live deliberately, to front only the essential facts of life, and see if I could not learn what life had to teach, and not, when I came to die, discover that I had not lived." I choose to live. Now, this body of mine, this old animal, is sixty-seven years old.

When we approach Lute Island by boat and gaze at it from a distance, a dollop of rock and green rising out of the sea, I am acutely aware that it will last far longer than I will. A hundred years from now, after I'm gone, many of these spruce and cedars will still be here. And the wind going through them will sound like a distant waterfall. The curve of the land will be the same as it is now. The paths that I wander may still be here, although probably covered with new vegetation. The rocks and ledges on the shore will be here, including a particular ledge I'm quite fond of, shaped like the knuckled back of a large animal. Sometimes, I sit on that ledge (more sitting) and wonder if it will remember me. Even my house might still be here, or at least the concrete posts of its footing, crumbling in the salt air. But eventually, of course, even this island will shift and change and dissolve. In geologic time, there may be no trace of Lute Island. Twenty-five thousand years ago, it didn't exist. Maine and most of North America were covered with ice, thousands of feet thick. Two hundred and fifty million years ago, the Atlantic Ocean didn't exist. Europe, Africa, and North America were joined together in a single landmass. Nothing persists in the material world. All of it changes and passes away.

Belief in many of the Absolutes—permanence, constancy, immortality, the soul, even God—must surely be driven in part by the knowledge of our personal mortality. I wish I could believe in life after death, as do most of the world's people. Or the soul. "Such harmony is in immortal souls," said Lorenzo in *The Merchant of Venice*. When a Jew leaves the world, the last prayer entrusts his or her soul to God for eternal life. My southern rabbi, a learned and charismatic man named Micah Greenstein, tells me that the afterlife is the ultimate reconnection with God. Micah, who is sweetly tolerant of my skepticism, explains to me that one of the Hebrew terms for God is *el maleh rachamim,* which means "God full of the womb." Thus in returning to

God after death, we return to the womb. I am struck by the similarity of this notion to Rolland's description of the "oceanic feeling," which Freud compared to the experience of total union between infant and mother. I wish I believed. I would have Lute Island for all of eternity, as well as my self and the selves of my loved ones and friends. So many questions and doubts would have possible answers.

To my knowledge, the most elaborate and detailed description of life after death is to be found in Tibetan Buddhism. I first learned about these beliefs during a Buddhist retreat in Wisconsin several years ago. By day, I sat very still and meditated with a dozen people in a domed circular room with huge windows looking out at the trees. By night, I cloistered myself in my small room and read *The Tibetan Book of the Dead*. According to Buddhist tradition, we pass through many cycles of life, death, and rebirth until we attain full enlighten- ment, at which point we enter the eternal and divine state of Nirvana. Between each death and rebirth, we inhabit an intermediate state, called *bardo,* in which our consciousness exists but is not attached to a physical body. Buddhist texts describe the various stages of *bardo* in intricate detail. Although conventional time and space would not exist within *bardo,* the time between each death and rebirth in the physical world is thought to be about forty-nine days for the average person.

Those fortunates who are able to attain enlightenment may exit the netherworld of *bardo* sooner.

I accept the idea that science and the scientific view of the world may not encompass all of existence. There could be realms of being outside the physical universe. We simply don't know what we don't know. And by definition, we would not be able to prove or disprove such nonphysical realms by methods or means within the physical universe. Science has been astoundingly successful at understanding the physical world. But science is restricted to that world. Spiritual realms, as I understand them, would not necessarily be logical or self-consistent or even comprehensible. But again, we don't know what we don't know. In my view, prominent scientists who use scientific arguments to attempt to disprove the existence of God are missing the point.

Still, I ask for some kind of evidence for all things I believe—even if it is evidence from a personal or transcendent experience. And I insist on evidence for any statements that concern the physical world. I do wonder how the Tibetan Buddhists arrived at such an ornate description of existence between death and rebirth. Have *bardo* travelers returned to our world and reported? Some claim they have. Some lamas do not make such claims themselves but are considered by others to have gone on the journey and returned. Putting aside the nonphysicality of *bardo*, how could these reports be convincing to the great majority of us who have not experienced them? In the end, you either believe or you don't.

At the conclusion of my ten-day Buddhist retreat in Wisconsin, I felt surprisingly refreshed, as when awakening from a long and restful sleep, but I had not altered my views of life after death. As I lie in my hammock now on this late afternoon in August, I can feel the seconds ticking away to my end, and I believe it to be a final end. But that finality does not diminish the grandeur of life. As the seconds tick by, I breathe one breath at a time. I inhale, I exhale. These spruces and cedars I cherish and know, the wind, the sweet scent of moist and dark soil—these are my small sense of enlightenment, my past life and present life and future life all in one moment.

The immortal soul of Judaism and Christianity and the disembodied consciousness that floats in the *bardo* of Buddhism, if they exist, could not be material. All material things—electrons and quarks, atoms and molecules, and the energy that makes and is made by these particles—are part of the physical world. And, according to the view of science, all material is governed by laws. Immortality is not allowed by the laws. Disembodied existence is not allowed by the laws. Insubstantiality is not allowed by the laws. How does a devotee of the laws of nature, a materialist like myself, think of his own personal death? For this particular materialist, I believe that the distinction between life and death may be overrated. I have come to believe that death occurs gradually, through the diminishing of consciousness.

Let me explain. According to the scientific view, we are made of material atoms, and nothing but material atoms. To be precise, the average human being consists of about 7×10^{27} atoms (seven thousand trillion trillion atoms)—65% oxygen, 18% carbon, 10% hydrogen, 3% nitrogen, 1.4% calcium, 1.1% phosphorous, and a smattering of 54 other chemical elements. The totality of our tissues and muscles and organs is composed of these atoms. And, according to the scientific view, there is nothing else. To a vast cosmic being, each of us human beings would appear to be an assemblage of atoms, humming with our various electrical and chemical energies. To be sure, it is a special assemblage. A rock does not behave like a person. But the mental sensations we experience as consciousness and thought, according to science, are purely material consequences of the purely material electrical and chemical interactions between neurons, which in turn are simply assemblages of atoms. And when we die, this special assemblage disassembles. The total number of atoms in our body at our last breath remains constant. The atoms remain, only scattered about.

Particularly special in these considerations is the

brain. In the view of science, the brain is where our self-awareness originates, our memories are stored, our elusive ego and "I-ness" are formed. Neuroscientists have studied the brain in great detail. Much is known. Much remains unknown. But the materiality of the organ is not in doubt. There is good evidence that the processing and storage of information is done by the brain cells called neurons. There are about a hundred billion neurons in the average human brain, and each neuron is connected by long filaments to between a thousand and ten thousand other neurons. The electrical and chemical components of these neurons are largely understood. The manner in which electrical signals are created and fly through the fibers of a neuron, then generate chemical flows at the juncture between one neuron and the next, then start up the electrical signal again in the next neuron is understood in quantitative detail. The creation of long-term memory, upon which so much of our self-identity seems to be based, is accomplished by the material generation of new connections between neurons and the strengthening of existing connections, all caused by specific proteins.

Despite the known material nature of the brain, the sensation of *consciousness*—of ego, of "I-ness"—is so powerful and compelling, so fundamental to our being and yet so difficult to describe, that we endow ourselves and other human beings with a mystical quality, some magnificent and nonmaterial essence that blooms far larger than any collection of atoms. To some, that mystical thing is the soul. To some it is the Self. To others, it is consciousness.

The soul, as commonly understood, we cannot discuss scientifically. Not so with consciousness, and the closely related Self. Isn't the experience of consciousness and Self an *illusion* caused by those trillions of neuronal connections and electrical and chemical flows? If you don't like the word "illusion," then you can stick with the sensation itself. You can say that what we call the Self is a name we give to the mental sensation of certain electrical and chemical flows in our neurons. That sensation is rooted in the material brain. And I do not mean to diminish the brain in any way by affirming its materiality. The human brain is capable of all of the wondrous feats of imagination and self-reflection and thought that we ascribe to our highest existence. But I do claim that it's all atoms and molecules. If our giant cosmic being examined a human being in detail, He/She/It would see fluids flowing, sodium and potassium gates opening and closing as electricity races through nerve cells, acetylcholine molecules migrating between synapses. But He/She/It would not find a Self. The Self and consciousness, I think, are names we give to the *sensations* produced by all of those electrical and chemical flows. Furthermore, there doesn't even seem to be an "executive branch" of the brain, a central location that executes decisions based on the inputs from other areas of the brain. Instead, neuroscience suggests that the cognitive functions are *distributed* throughout all of the neural circuits.

If someone began disassembling my brain one neuron at a time, depending on where the process began I might first lose a few motor skills, then some memories, then perhaps the ability to find particular words to make sentences, the ability to recognize faces, the ability to know where I was. During this slow taking apart of my brain, I would become more and more disoriented. Everything I associate with my ego and Self would gradually dissolve away into a bog of confusion and minimal existence. The doctors in their blue and green scrub suits could drop the removed neurons, one by one, into a metal bowl. Each a tiny gray gelatinous blob. Stringy with the axons and dendrites. Soft, so you would not hear the little thuds as each plopped in the bowl.

Likewise, the same doctors in their blue and green scrub suits could create consciousness by building a brain from scratch, one neuron at a time, delicately arranging the connections between neurons. The doctors might connect some of the neurons to a device that monitored their combined electrical activity. Neuron by neuron, connection by connection. At first, there would be simply noise. But at some point, presumably, a change would occur, a coherent signal, an unusual hum, that would translate, roughly, into "Yikes, something is messing with *me*."

If we conceive of death as nothingness, we cannot imagine it. But if we conceive of death as the complete loss of consciousness, a view supported by the understanding of the body as an arrangement of material atoms, then we *approach* death in gradual stages as consciousness fades and dissolves. The distinction between life and death would no longer be an all-or-nothing proposition.

The neuroscientist Antonio Damasio has defined different levels of consciousness. The lowest level, which he calls the "protoself," is related to an organism's ability to carry out the most basic processes of life, but nothing else. An amoeba has a protoself. I would not associate this level of existence with consciousness. Almost certainly, thought and self-awareness require a minimum number of neurons, well beyond the stuff of an amoeba. Next comes "core consciousness." It is self-awareness and the ability to think and reason in the present moment, but without memories extending earlier than a few minutes into the past. Such an organism, far above an amoeba, might be able to have an understanding of the world around it and its place in that world, but it would exist only in the present.

People with certain brain disorders have only core consciousness. They cannot form new memories that last more than a few minutes. They cannot remember what happened to them in the past, except for isolated periods. For the most part, they cannot recall past personal relationships or the people they loved and who loved them. They cannot make plans for the future. They are trapped in the moment.

The highest level of consciousness is "extended consciousness," which all healthy human beings possess. Here, we can remember most of our past life as well as function completely in the present. We can remember our view of the world based on past experiences, we can remember our value system as grounded in those experiences, we can remember what we like and don't like, places we've been and people we've met. Self-identity, as most psychologists understand it, probably requires extended consciousness, that is, long-term memory. These are complex issues, not fully understood.

The slow dismantling of a human brain, whether by my imaginary doctors in their scrub suits or by the deterioration of the brain in neural disease, might proceed from extended consciousness to core consciousness to the protoself. Or perhaps it might proceed in a less orderly manner, by removing chunks of extended consciousness and core consciousness here and there until nothing is left but the protoself. However it proceeds,

one begins with full consciousness and ends with an amoeba-like existence, alive only in the biologists' formal definition of the word. One begins with a full life and ends with death, or the equivalent of death. And this process can happen gradually, so that there may be some awareness of the increasing loss of awareness.

Personal accounts of early dementia provide the best knowledge we have of approaching death in this manner. In the early stages of dementia, enough of the mind remains to understand and articulate what is happening. In later stages, the reporter has slipped and disappeared into the abyss of confusion. Somewhere in that intermediate nether zone, the sense of self dissolves and is gone. It's a grim subject. One account, by "Leo from Tasmania," goes as follows:

> When leaving my doctor's surgery one day I could not find my car. I did not know where I was. Eventually, I was able to make my way home . . . Independence has featured strongly throughout my life. I now depend on my wife, Ellie, to oversee any decisions I make. I find this very difficult. Timing and how you say things is very important, but my sense of timing is now lost. Say it now or forget it. People have disappeared from my life. It is like going through a divorce. I fear making a fool of myself.

Another person, Ted, describes his life with dementia in the following way:

My interaction with our grandchildren is different now as they know that I am different, but are not sure just what the problem is . . . I am no longer comfortable taking them on road trips to various parks as I did in the past for fear that I will have a memory lapse and forget them or somehow endanger their safety. Now I am reluctant to go places alone, and my wife has retired early to be with me at all times. . . . Additionally, I have few new memories to speak of as recent events just do not stay in my mind on a consistent basis. As a result, I feel that I am robbed of any future because while I will live in it, I will be unlikely to remember it. Finally, I could not write this without the help of my wife. . . . My wife is now my memory, and I can write this only with her ability to remember parts of my life that I am getting fuzzy about. In some cases, I can forget what I was trying to say while typing the sentence, so I tell her what I was getting at, so she can help if it fades out as I type it.

Some of my own loved ones have gone through various forms of dementia. Many of us will not suffer this depressing approach to death, and I prefer not to think

of it myself. But consciousness and its loss are part of my musings today on the boundary between life and death. For a person who does not believe in the afterlife, consciousness is the subject of interest. For a materialist, death is the name that we give to a collection of atoms that once had the special arrangement of a functioning neuronal network and now no longer does so.

From a scientific point of view, I cannot believe anything other than what I have laid out above. But I am not satisfied with that picture. In my mind, I can still see my mother dancing to the bossa nova as she often did, giving her hips a jaunty shake with the beat. I still can hear my father tell his Cooshmaker joke—"It went coosh, and I can have another one up in fifteen minutes"—and his comment years ago when the family went on a sailing trip together: "This is what happiness is." I often wonder: Where are they now, my deceased mother and father? I know the materialist explanation, but that does nothing to relieve my longing for them, or the impossible truth that they do not exist.

Long ago, the Roman philosopher and poet Lucretius, whom I have quoted before, said that we should not fear death, especially life after death, because we are made completely of atoms (which Lucretius called "first bodies"), and when we die those atoms are simply dispersed.

Since I have shown it [spirit] to be delicate
and composed of minute particles and elements
much smaller than the flowing liquid of water or
cloud or smoke . . . therefore, since, when vessels are shattered, you perceive the water flowing out on all sides and the liquid dispersing, and
since mist and smoke disperse abroad into the air,
believe that that spirit also is spread abroad and
passes away far more quickly, and is more speed-
ily dissolved into first bodies [atoms] . . . Therefore death is nothing to us . . . For, if by chance
anyone is to have misery and pain in the future,
he must himself also exist then in that time to be
miserable.

A contemporary writer once said that the worst part
of death is not the actual end, at which point we no
longer have feeling or thought, but the physical experience of aging—the *approach* to the end. In aging,
we slowly lose our mental and physical capacities and
must contend with an increasing chorus of aches and
pains, diminishing eyesight and hearing, sore joints, the
fading of memory. And, for some, the gradual deterioration of consciousness.

I have a confession to make. I am alive. And I am
enjoying it, aching joints and all. Despite my belief that
I am only a collection of atoms, that my awareness
is passing away neuron by neuron, I am content with

the *illusion* of life. I'll take it. And I find a pleasure in knowing that a hundred years from now, even a thousand years from now, some of my atoms will remain on Lute Island. Those atoms will not know where they came from, but they will have been mine. Some of them will once have been part of the memory of my mother dancing the bossa nova. Some will once have been part of the memory of the vinegary smell of my first apartment. Some will once have been part of my hand. If I could label each of my atoms at this moment as I lie in this hammock, imprint each with my Social Security number, someone could follow them for the next thousand years as they floated in air, mixed with the soil, became parts of particular plants and trees, dissolved in the ocean and then floated again to the air. Some will undoubtedly become parts of other people, particular people. Some will become parts of other lives, other memories. And some, after that long journey, will return to Lute Island.

Certainty

The man sits in a drafty room, writing by the light of a single oil lamp. One hand rests on an open Bible, the other on his manuscript, inking the words with a thin feather quill. In these words, the man is confessing his sins and asking for God's grace. Let us put the year at AD 400. We will imagine the scene as depicted in Caravaggio's painting. Our man wears the robes of the clergy. Partly balding but with a full beard, his cheeks glow in the lamplight. His hands are delicate, almost feminine. Behind him, standing upright on a bookshelf, is the pointed miter of a bishop, its red ribbons casually draped over the books as if in repose for the night. In fact, our man is the Bishop of Hippo. He presides over this Roman city on the north coast of Africa, a city filled with stone churches and new Christians. Tonight, he works alone in a small room of the monastery. In the dim light, the Bishop of Hippo seems enveloped in quiet, peace, and serenity. At the same time, his mind burns with ideas. He appears deep in thought over the project at hand, which will become one of the most influential tomes in Christian theology.

Our man is Aurelius Augustinus Hipponensis. Eventually, he will be known as Saint Augustine, sometimes as Blessed Augustine, sometimes as the Doctor of Grace. He might also be called the Monarch of Certainty. For few people have so forcefully expressed the certainty of God's grace, of the immortality of the soul, of absolute truth, of absolute morality. Aristotle allowed for degrees of certainty, especially in moral affairs. Augustine made no such allowance. Augustine's certainties were absolute. For example, Augustine believed that lying was *always* a sin, regardless of the circumstances—even if a lie was needed to shield the life of one's own child. Augustine's "City of God," his proposed spiritual universe in parallel to the material universe, floated in a divine ether of Absolutes. On upright behavior, truth, good and evil, God, predestination, Augustine gave no quarter. He made no compromise. He began with the certainty that he was alive—a certainty arrived at because he could think. On this reasoning, he preceded Descartes by more than a thousand years. From that certainty, he moved to the certainty of God and the authenticity of the Holy Scriptures as the word of God. (Augustine's writings are filled with quotations from the Scriptures, as a physicist's exposition bristles with irrefutable equations.) From Scriptures, the Monarch of Certainty marched on to absolute truth, deriving all things from God, "of Him out of Whose mouth nothing false proceeds." From there to purity and absolute morality: "pious, true, holy chastity is not otherwise than of the truth." For Augustine, purity was a special challenge. Acknowledging on many pages of the *Confessions* his own sinful ways, Augustine fiercely supported the idea of original sin, yoked to all human beings after the fall of Adam and Eve. Furthermore, and to the dismay of many in the Church, Augustine declared that only a few people could be absolved of that sin, that those few only by the grace of God, and that only God knew which people would be saved. In other words, the eternal destiny of each human being is already ordained (predestined), and for most of us the news is not good. The Buddha also claimed certainty about particular aspects of existence—for example, that the cosmos will cycle an infinite number of times.

I've been reading Augustine on the island in Maine. I'm impressed by his faith and sincerity as much as by his intelligence. His writing reminds me of Rabindranath Tagore's *Gitanjali*. Tagore expresses his devotion to God in a hundred short poems, each poem refracting God's presence in a different light. In my island study, I keep a few of my most precious books. Among them is the collection called the Harvard Classics, edited by

Charles W. Eliot. Professor Eliot was president of Harvard from 1869 to 1909. In his last year, he put together what came to be called "Dr. Eliot's Five-Foot Shelf of Books," fifty volumes of the greatest Western literature and thought as judged by Dr. Eliot. Volume 7 contains the *Confessions of St. Augustine*. The *Confessions* is just a tiny fraction of Augustine's leviathan production. He was the author of some eighty books, with titles such as *The City of God* (his most famous), *On the Immortality of the Soul, On Free Will, On Lying, On the Nature of Good, On Nature and Grace, On Patience,* and *On Adulterous Marriages*. He wrote numerous private letters to laypeople and bishops alike expounding his views of God's power and grace. Much of his thought leaped from his lips in his seemingly extemporaneous sermons, eight thousand of them according to some estimates. A superb orator as well as brilliant thinker, Augustine at age thirty was appointed professor of rhetoric at the imperial court of Milan, considered the highest academic post in the Latin world at that time. A half-dozen years later, soon after his conversion to Christianity, Augustine crossed the Mediterranean to Hippo.

I want to understand how Augustine climbed to the apex of Certainty. What crampons did he employ? None from the physical world. The physical world, and especially the world of humanity, churns with uncertainties. Friendships hold fast for decades and then rip without warning. Nice people suddenly do unpleasant things, and vice versa. Who can predict with certainty whether two particular people will fall in love? That we will all die is a certainty, but the particular date cannot be foretold. Disease. Jobs and careers. Marriages. Weather. Asteroids. Ice ages. The zigzagging path of one leaf as it drops from a tree. Fortunately, we do not need complete certainty to live in the world. We require only a modicum of certainty. We can live with approximate certainty, and approximate certainty is what we are given.

Absolute and complete certainty, Augustine's certainty, was and is a different kind of thing. Augustine's God was not a probability. The limits of God's power were not debatable. Augustine's ideal moral behavior had no ambiguity or qualifications. Those who were not saved were doomed to certain and eternal damnation in Hell. What other certainties could Augustine have imagined in his day? Euclid's geometry did indeed prove theorems with absolute certainty, provided one accepted the initial axioms and assumptions. But mathematics was unique. Mathematics was a formal system of logic: "All dogs pee in the road. Philus is a dog. Therefore, Philus pees in the road." Augustine's cosmos of absolute certainties was not a formal system, it was not necessarily logical, and it was not a tautology. Nor was it drawn from the physical world. Augustine's absolute certainty was a mental construction. It lived in the

City of God. It was a beautiful abstraction. Science in AD 400 was also a beautiful abstraction. Although cause-and-effect relations were well understood, the method of testing hypotheses against physical experiment was several centuries off in the future. In AD 400, physicians carried out diagnoses and treatments according to the positions of the planets and stars. Aristotle's physics and astronomy were full of well-argued decrees, such as the (incorrect) statement that heavier bodies fall faster than light ones, but Aristotle was not an experimentalist. He reached his conclusions from an armchair. And he didn't check.

According to his *Confessions,* Augustine was visiting a friend when he first heard the word of God spoken to him. He was thirty-one years old at the time, regretting the wild and lustful escapades of his youth. "Soul-sick was I, and tormented, accusing myself much more severely than my wont, rolling and turning me in my chain," he wrote, when he heard from a nearby house the voice of a child chanting "Take up and read; Take up and read." Interpreting these words as the direct command of God, Augustine immediately opened the Bible to a random page and read the following passage from Romans: "Not in rioting and drunkenness, not in chambering [illicit sex] and wantonness, not in strife and envying; but put ye on the Lord Jesus Christ, and make not provision for the flesh." As Augustine then recalled, "Instantly at the end of this sentence, by a light as it were of serenity infused into my heart, all the darkness of doubt vanished away."

Augustine had found Certainty.

Certainty, like permanence and immortality, is one of those conditions we long for despite a great deal of evidence to the contrary. Certainty often confers control. And we badly want control in this strange cosmos we find ourselves in. In his classic study *The Golden Bough,* anthropologist James George Frazer discusses how primitive people developed magic so that they could control a world filled with the uncertainties of lightning and storms and vicious animals. The Bantus in Botswana burn the stomach of an ox in the evening because they think the black smoke will gather the clouds and cause the rain to come. Certainty offers us safety, stability, reliability, predictability, rules for behavior. If I am completely certain that it is unethical to harm other people's careers in order to advance my own, that certainty provides a clear and constant guide for how to conduct my professional life. Augustine's absolute certainty about theological and ethical matters may well have been an extension of a psychological and physical desire for certainty.

And then there's the practical. Certainty, whether real or imagined, permits us to predict the future, at least in the physical world. And successful prognostication confers survival advantage. If a heavy coconut has just snapped off a tree I am standing beneath, it is beneficial for me to be able to estimate its future trajectory and know whether I should step to the left or the right to avoid getting my skull crushed. Likewise, the ability to predict the time of nightfall, so that I can retire safely to my cave, or the ability to predict the seasons, so that I can plant and harvest, or the ability to predict rain (by burning an ox stomach or trusting the weather reports), so that I can plan outings.

If a preference for certainty has been wired into our brains over the millennia as a tool for survival, it follows that uncertainty should cause stress and discomfort. Researchers at the California Institute of Technology and the University of Iowa Medical School have done experiments showing that there are entire neural networks in the brain dedicated to evaluating the level of uncertainty in the decision process. In particular, the scientists monitored the brain activity of subjects who were required to make decisions after being given varying amounts of information related to those decisions. The researchers found that the greater the uncertainty (in the form of less information), the greater the electrical activity in the amygdala. The amygdala, the most primitive part of the brain and what one might call the "ancient brain," is known to play a primary role in memory, decision making, and the emotions. For example, the fear response occurs in the amygdala. It is no wonder that we feel anxiety in the face of uncertainty, contentment and calm with certainty. These findings support the idea that our craving for certainty originates in a deep part of our psyche and is part of our most ancient DNA.

Augustine's absolute certainties, as I understand them, are necessarily a matter of faith. As mentioned earlier, I have come to appreciate that science also operates on faith. It is faith that the physical world is a territory of order and logic. And that faith derives from probing the external world. By contrast, the *Confessions* makes it clear that Augustine arrived at his faith not through external evidence and rational analysis but through his personal passion and introspection. In fact, as the great philosopher and theological thinker Søren Kierkegaard has emphasized, religious faith may actually be *antithetical* to reason and rational analysis. In Kierkegaard's view, conclusions based on observation of the external world are never certain. We cannot fully commit to these conclusions because they are being constantly revised (like science). However, we can commit to our personal inner passions because they are not subject to the accident and swerve of the outer world. We know them to be true because we directly experience them.

Kierkegaard does indeed acknowledge that science

and mathematics seek objective truths, but those truths are uninteresting, what he calls "indifferent truths." To be interesting, to have value and meaning, says Kierkegaard, a truth must be internalized, merged with our bloodstream, imbued with personal passion, with our humanity. (Recall the similar version of truth espoused by Tagore in his conversation with Einstein.) In that inner merging, truth becomes closely aligned with faith. Writes Kierkegaard: "When subjectivity is truth, the definition of truth must include an expression of the antithesis of objectivity . . . another way of saying faith." Finally, we cannot know God in any objective sense. So we must go inward to the subjective to know God. "If I can grasp God objectively," says Kierkegaard, "then I do not have faith, but just because I cannot do this, I must have faith."

Science, while it does not possess any absolute certainties, is a continuous search for certainty, in the form of completely accurate predictions of the future, the "final theory." And a predictable universe is a universe of determinism and causality. The great theoretical physicist Max Planck once said that "the assumption of absolute determinism is a necessary basis for every scientific investigation." The idea is that the future is completely determined by the past through the action of the laws of nature.

The deterministic universe has sometimes been likened to a giant clock, a totally mechanical system. Wind up the clock and its parts move according to the springs and gears within. As far as science is concerned, an all-knowing Intelligence may or may not have designed the clock. An all-knowing Intelligence may or may not have wound up the clock. But once set in motion, the hands of the clock turn inexorably and autonomously according to the mechanism within.

One of the leading figures responsible for the clockwork conception of the universe was Isaac Newton. In Newton's view, the physical world is composed entirely of masses and forces. Take any particle like an atom or a coffee mug, specify its initial location and velocity and the forces acting on it, and its position at any time in the future is completely determined. More generally, the future of the entire universe is completely determined by its past, via the laws of nature. Or, as the French scholar Pierre-Simon Laplace put it a century after Newton's *Principia,*

We ought then to regard the present state of the universe as the effect of its anterior [past] state and as the cause of the one which is to follow. Given for one instant an intelligence which could comprehend all the forces by which nature is animated and the respective situation of the [objects] that compose it . . . nothing would be uncertain and the future, as the past, would be present to its eyes.

The clockwork universe is a *causal* universe. Every event is caused by a previous and distinct event, which was caused by a previous event, and so on, in a long chain of causes and effects extending back to the beginning of time. Cause and effect. Cause and effect. Cause and effect.

As I sit here at my desk pondering Augustine's predestination and Newton's clockwork universe, what haunts me most is the question of whether I am a piece of the clockwork myself. True, I do not fully know the mechanism of that clock—no one does—but I believe that the mechanism exists. Are all of my decisions and actions determined by prior events with complete certainty, or do I have the freedom to act in this moment in some unpredictable way—unknown to God, unknown to the hexagrams of the I Ching, unknown to the universe? In other words, am I a cog in the machine? This question, of course, is one of the old chestnuts of philosophy, under the heading of Free Will versus Determinism.

Surprisingly, Augustine believed that we have free will even though God knows in advance how each decision will be made. As the Bishop of Hippo commented to his hypothetical sparring partner in *On Free Will*,

> Unless I am mistaken, you would not directly compel [a] man to sin, though you knew beforehand that he was going to sin. Nor does your prescience in itself compel him to sin even though he was certainly going to sin, as we must assume if you have real prescience. So there is no contradiction here. Simply you know beforehand what another is going to do with his own will. Similarly, God compels no man to sin, though he sees beforehand those who are going to sin by their own will.

Augustine, being a firm believer in absolute good and evil, and in sin and redemption (for a few), required free will for sin to have meaning. In order to sin, we must have a free choice between good and evil at each point of decision. Personally, I am not persuaded by this argument. I cannot begin with the assumption that the ability to sin is an essential element of the human condition. So, I must look elsewhere than Augustine to answer my question of whether I am an independent decider, or a machine.

Science, it would seem, argues that I am indeed a machine. That I am so many molecular gears and wheels, currents and chemical flows, inexorably grinding on by cause and effect, cause and effect. Science argues that my actions are completely determined by the state of all masses and forces in the past.

In the early twentieth century, this clockwork view of the universe was challenged by an area of physics called quantum mechanics. Quantum mechanics does not repudiate the Central Doctrine of Science. Rather, it says that it is impossible to know with complete certainty the state of all masses and forces at any moment in time. It is impossible to know with complete certainty the position and speed of even a single atom. This uncertainty, articulated by the physicist Werner Heisenberg in 1927, is not a matter of the approximate nature of measurement. It is a fundamental property of nature as we know it. Even with infinitely precise measuring devices, we cannot determine the position and speed of a single particle with complete certainty. And without that precise knowledge, I cannot say with complete certainty where my particle will be in the future—even if I know all the forces acting on it and all the laws of nature with complete certainty. The validity of quantum physics and the Heisenberg Uncertainty Principle has been well established in laboratories all over the world. Quantum physics, in fact, lies at the heart of the workings of transistors, computer chips, and many technological devices of our age.

That said, the uncertainties inherent in quantum reality rapidly diminish in importance as we move from single elementary particles like electrons and photons to objects comprising a huge number of particles, like neurons and other biological cells. Essentially, the uncertainties in the initial conditions of individual particles get averaged out in a large aggregation of particles, just as the random speeding up and slowing of individual cars in a long line of heavy traffic averages out to a constant speed at the end of the line. Turning to the question at hand—decisions made by a human mind—modern biology suggests that the fundamental operative unit of the brain is the neuron. It is reasonable to assume that the "initial condition" of a neuron is specified by its internal chemical configuration, by which other neurons its axons connect to, and by the chemical strength of each such connection. Since the axons themselves and the chemical messengers released by axons involve many millions of molecules, it is hard to believe that quantum uncertainty could have any significant effect on the specification of the initial state of a neuron. Thus, even though the Heisenberg Uncertainty Principle says that in principle we cannot know the *exact* present condition of a single neuron or a system of neurons and therefore cannot precisely predict their future, the uncertainty gets averaged to insignificance for all practical purposes. So I seem to have argued myself back to the view that I am a machine.

But the question of free will versus determinism is

subtler than these reductions. Because even if I am a machine, and even if a supercomputer connected to my brain could predict all of my future decisions with complete certainty, that does not mean that I *myself* can predict my own future actions and decisions. Such predictions require another layer of thought: *awareness of my own mind,* that is, awareness of a mind that is aware of its own workings. So, now I am a self-referencing machine. And that leads to serious problems, as in all self-referencing propositions. To be aware of my own mind, I must be aware of a mind that is self-aware, and that envisioned mind must be aware of a mind that is self-aware, ad infinitum, and the problem dissolves into an infinite and unending regression, like two mirrors pointing toward each other, each mirror showing its own reflection, which contains another mirror showing a smaller image of itself, which contains another mirror showing an even smaller image of itself, ad infinitum.

On further reflection, even these extended considerations may be missing a big point. The notion of free will requires a Willer, an "I" making the decisions, a pilot in the cockpit. And as I have written before in these pages, I believe that the "I" is an illusion. I believe that there is no I, no Self. In my view, and the view of many biologists, the powerful feeling of consciousness and Self is just a name we give to the mental sensation of a hundred billion neurons sending electrical and chemical impulses back and forth in our skulls. In fact, the first person "I" that I have so frequently used with abandon in these pages is a fraud, a construct, an illusion. In this case, the question of free will versus determinism may not be easily defined. Clearly, there is some mechanism at work to allow the creation of ideas and words and keystrokes on "my" computer. But it is a complex mechanism, with no single neuron or group of neurons wearing the admiral's hat. For convenience, I will bow to convention and continue to act and write as if an "I." It's too cumbersome to replace "I" by "the electrical and chemical flows that are being exchanged between a hundred billion odd neurons in the particular skull known as Alan Lightman in the year 2016." Mulling over these new ideas, "I" now consider "myself" a machine even more than before.

I will admit that I'm not feeling cheerful after these ruminations. I'm going to get up from my desk, at which I've been sitting far too long, and take myself for a walk down a winding path that meanders to the northern tip of Lute Island, where there's a weathered bench I know well. The path curves this way and that, closely overhung on both sides by mountain laurel, bayberry bushes, and the branches of spruce trees. At any point along the way, I cannot see more than six feet ahead, so that I feel as if I'm walking through a narrow green tunnel. But I happen to know there's a lichen-covered bench waiting at the end. As my feet fall softly on a carpet of moss, I can hear the constant music of the island, the breathing of the ocean, the songs of unseen birds. Listen. In my head, I hear the opening movement of Vivaldi's "Spring," his evocation of birds. Always music. Now I'm sitting on the bench. The ocean lies thirty feet away. Looking across the bay, I can see a few faint houses on the mainland, nestled in the woods. In the distance, I can see the masts of sailboats moored at a marina.

So, my hundred-billion-odd neurons are sitting here, thinking thoughts that are the direct result of particular electrical and chemical flows, which in turn have been caused by previous electrical and chemical flows, and so on, in a deterministic chain of causes and effects extending back to $t = 0$. What is this poor machine to do? Should it stop entertaining the notions of Self, of "I-ness," of agency, of decision making, of good and bad behavior, of meaning itself? One thing it (I) knows is that it feels pleasure and pain. And when I (it) talk about pleasure and pain, I do not mean merely physical pleasure and pain. Like the ancient Epicureans, I mean all forms of pleasure and pain: physical, intellectual, artistic, moral, philosophical, and so on. Like Augustine's certainty of being alive, from which all other certainties derived, I am certain that I feel pleasure and pain. "I feel, therefore I am." And here is the point I have finally attained. Since I cannot escape these sensations, machine or not, I might as well live in such a way as to maximize my pleasure and minimize my pain. Accordingly, I will try to eat delicious food, try to support my family, try to create beautiful things, and try to help those less fortunate than myself, because those activities bring me pleasure. Likewise, I will try to avoid eating uncooked eggplant, avoid leading a dull life, avoid personal anarchy, avoid hurting others, because those activities bring me pain. That is how I should live. A number of thinkers deeper than I have come to these same conclusions via other routes.

Looking out on the water, I see a haze in the air, a softening of edges. I see ocean and sky and a few seagulls sailing through space. I think I've decided to sit here a while longer.

Origins

On Wednesday, February 11, 1931, Albert Einstein met for more than an hour with a small group of American scientists in the cozy library of the Mount Wilson Observatory, near Pasadena, California. The subject was cosmology. For years, Einstein had insisted, like Aristotle and Newton before him, that the universe was a magnificent and immortal edifice, fixed and unchanging for all of eternity. When a prominent Belgian scientist proposed in 1927 that the universe was growing in size like an inflating balloon, Einstein pronounced the idea "abominable." Recently, however, the great physicist had been confronted with telescopic evidence that the distant galaxies were in flight. They were in motion. In his thick German accent, he told the surrounding men in their suits and ties that the observed motion of the galaxies "has smashed my old construction like a hammer blow." Then he swung down his hand to emphasize the point. What rose in the shards of that hammer blow was the Big Bang cosmology:

the universe is not static and everlasting; rather, it is expanding. Playing that movie backward in time, the universe "began" some 14 billion years ago in a state of extremely high density and has been growing and thinning out ever since. Was that "beginning" also the beginning of time?

I have been thinking of Augustine. Saint Augustine had an idea about time—time born in the sensorium of God. For the Doctor of Grace, time like all things was created by God, presumably during His creation of heaven and earth. When impertinent folks asked Augustine what God was doing *before* that monumental act, he replied that God "art the Creator of all times . . . nor could times pass by before [God] madest those times." Prior to heaven, and earth, Augustine said, "there was no 'then,' when there was no time." There was no before. Could Saint Augustine, or anyone, imagine existence without time? Alternatively, could anyone imagine eternity?

Permanence, immortality, eternity—these are all Absolutes. Of course, our main concern with these notions is personal: What will happen to us after *we* die? Will that be the final end, or will we continue to exist in some eternal realm? Although few people would argue that such a realm is physical, one is compelled to ask the same questions of the physical world.

If the universe began some 14 billion years in the past, as scientists now think, was there anything in existence *before*? And is that even an answerable question? If not, the origin of the universe might be the dot where the Absolute and the Relative converge.

All questions of origins are difficult to grasp, starting with one's personal origins. Here's what we see in a time-lapse video of an actual human cell just after sperm and egg meet: On day 1, right after fertilization, there's only a single cell with two nuclei, male and female. The cell is roughly spherical, whitish in color with gray patches here and there, like the surface of the moon as seen by a low-power telescope. On day 2, there are two cells, each with its own nucleus, which then divide to make four cells. On day 3, another division makes eight cells. On day 4, there are many divisions. Could each of us have originated from *that*? It seems impossible. And yet we are told that it's true.

Let's continue this time-lapse of the true but unfathomable, now going backward in time. From photographs, I can imagine the early lives of my parents before I was born, each of whom also miraculously originated in a tiny whitish thing ten times smaller than a mustard seed. Likewise, I can try to imagine my parents' parents, and *their* parents, backward in time, back and back through the generations. Very soon, the faces dissolve and I am sailing a sea of abstractions. Eventually, I come to some of my ancestors ten thousand years in the past, whose DNA remains in my body. I would love to sit in a fire-lit cave with those people for an hour and just acknowledge that we are connected in time, blood into blood. As unfathomable as it is, those people once lived; there's no doubt of it. I don't know their names, but they lived, those relatives of mine.

Permit me to go further. I can trace those ancient ancestors back to the first humans, and then to the first primates, and then to the first squishy things half–land animal and half-fish, and then to the one-celled amoebae swimming about in the primordial seas. And then to the gelatinous goo of the not-quite-alive, to the formation of earth's atmosphere, and then to the formation of earth itself by the slow condensation of gas and debris, and then to the star that created that gas. Scientific evidence tells me it all happened.

Continuing back, and strictly following the findings of science, all the stars in the sky were once nebulous clouds of gas. The further back in time, the hotter and denser. All the atoms were once too hot to hold their electrons. And before that, the nuclei of atoms were too hot to hold their constituent protons and neutrons. All the matter in the universe was once pure energy, and that energy, the stuff that made the entire observable universe today, was once crammed, churning and screaming, into a region smaller than a single atom. That was the Big Bang, the Origin of all origins. Or maybe not.

The most profound questions seem to have this fascinating aspect: Either they have no answer at all, or all possible answers seem impossible. So, here's one more profound question: Did anything exist *before* the Big Bang? Was the Big Bang the *beginning of time*? Or was there something before, some kind of eternal "meta-universe" that spawned our universe and possibly other universes? There are only two possible answers. Either reality and time existed throughout the infinite past (what one might call a meta-universe), or reality and time began at a finite moment in the past (the Big Bang of our universe). Believers in God may or may not take the meta-universe to be God. Infinite or finite, that's your choice. Either possibility is unsettling. Infinity cannot be comprehended by mortals, except in abstract mathematical terms. As for the other alternative, that there was somehow a beginning of time, a beginning of everything, we are faced with the question of what caused that beginning. How did reality and time emerge out of *nothing*? Only in the last hundred years has science been able to weigh in on these questions—after millennia of philosophical and theological speculation.

The oldest recorded story of Creation is the Sumerian Enuma Elish. It tells how in the Beginning, before sky or ground, there was only Apsu, the sweet waters, and Ti'amat, the salt waters. In time, these stretched into the giant ring of the horizon. "Then were created the gods in the midst of heaven. Lahmu and Lahamu were called into being . . . Ages increased." The ancient Babylonians and Sumerians based much of their civilization on the Tigris and Euphrates rivers, so it is natural that their genesis myth would involve water. By contrast, the ancient Egyptians, installed on another famous river, did not have a creation myth. In fact, they viewed time not as linear but as cyclical. The regular and predictable cycles of the Nile offered not only food security but also the vision of an eternal and cyclic universe, in balance, without beginning or end.

Anaximander (610–547 BC) proclaimed that there are an infinite number of worlds, which separate from the infinite (*ápeiros*), come into being, and then perish and are reabsorbed into the ageless infinite. The idea is not dissimilar to Buddhist cosmology, in which all beings and things come and go through cycles but end up joining some kind of eternity (Nirvana). Aristotle, too, argued that the universe had to be eternal, without beginning, because all matter, he claimed, comes from a prior substance.

Most other cultures have viewed the universe as having a beginning in time, its origin in the action of God. The Hindu Rig Veda hauntingly states:

If in the beginning there was neither Being nor Non-Being, neither air nor sky, what was there?

Who or what oversaw it? What was it when there was no darkness, light, life, or death? We can only say that there was the One, that which breathed of itself deep in the void, that which was heat and became desire and the germ of spirit.

Saint Thomas Aquinas incorporated most of Aristotelian thought into Christianity, but rejected the idea of an eternal universe in favor of an origin as described in the Bible. As did Moses Maimonides for Jewish thought. As did Muhammad for Islamic thought. From the Qur'an: "The Originator is He of the heavens and the earth: and when He wills a thing to be, He but says unto it, 'Be'—and it is."

As far as I know, all major religions that subscribe to a belief in God—including Judaism, Christianity, Islam, and Hinduism—believe that the universe was created by God at a *finite* time in the past. The one major contemporary religious tradition that does not incorporate God, Buddhism, holds that the universe has existed for all of eternity. Looked at another way, a universe with a beginning must have had a creation, either by a divine being or by quantum physics. But a universe that has existed forever needs neither.

In 1929, cosmological thinking changed radically. It moved from the realm of philosophy and theology into the province of science. During that year, the American astronomer Edwin Hubble discovered evidence that the universe is expanding. In particular, using the giant new telescope at Mount Wilson, California, Hubble found that the galaxies are all speeding away from each other, and that the speed of recession is proportional to their distance. In other words, a galaxy 20 million light years away is receding from us at twice the speed of a galaxy 10 million light years away. That result is exactly what you find if you paint dots on a balloon and begin blowing it up. From the vantage point of any one dot, all the other dots are moving away from it, and their recessional speed is directly proportional to their distance away. In this picture, there is no central dot. The view from any dot is the same. Play the movie backward, and the dots get closer together, until all the dots are on top of each other. That is the beginning, the "Big Bang," what cosmologists call $t = 0$. Recent cosmological measurements have refined the calculation of the age of the universe. It is believed with high confidence that the Big Bang occurred 13.8 billion years ago.

When I first read about the balloon metaphor in high school, I wondered about the center of the balloon. Wouldn't that be the center of the universe? But the cosmic balloon metaphor, like all metaphors, is not perfect. The metaphor applies only to the *surface* of the balloon. For purposes of illustration, space has been reduced by one dimension and exists only on the surface of the balloon. In the metaphor, the interior of the balloon does not correspond to anything in the physi-

cal universe. The real universe is three-dimensional, of course, but we are using a two-dimensional analogy to illustrate the way dots (or atoms or galaxies) move away from each in a space that is expanding. The main point here is that in a uniformly expanding space—whether two-dimensional, three-dimensional, or sixteen-dimensional—the distance between any two points increases, the speed of their separation is proportional to their distance apart, and the view from any point is the same. There isn't a center.

The Big Bang idea and the evidence supporting it have forced scientists to wrestle with the question of the origin of the universe. This issue did not have to be confronted in the static cosmologies of Aristotle and Newton and Einstein. Even after the work of Hubble, a number of prominent scientists found the question of cosmic origins sufficiently disturbing that they proposed alternative cosmologies to avoid a "beginning." For example, in the 1930s Professor Richard Tolman of the California Institute of Technology suggested the idea of an oscillating universe. In this picture, the universe expands (as it is doing now), reaches a point of maximum expansion, collapses to a tiny size, begins a new expansion, and repeats the process of expansion and contraction through an infinite number of cycles (in the Buddhist tradition). No origin need be consid-

ered. Tolman's oscillating model was in vogue from the 1940s to the mid-1960s. In a seminal paper in 1965, in which he predicted the existence of cosmic radio waves, the great Princeton University physicist Robert Dicke wrote that an oscillating universe, having existed for all time, "relieves us of the necessity of understanding the origin of matter at any finite time in the past."

In 1948, a group of restless young theoretical astrophysicists at Cambridge University, led by Fred Hoyle, came up with the "Steady State" model of cosmology. This model proposed that the universe, on average, does not change in time. The Steady State model reconciled itself with the observation of the outward motion of galaxies by postulating that new galaxies are continuously created throughout space to keep the density of galaxies constant. In the Steady State model, the universe, on average, has always looked as it does now. It was never denser in the past. It never had a beginning.

For various observational and theoretical reasons, both the oscillating universe idea and the Steady State idea have been ruled out. We are left with the Big Bang, and what may or may not have preceded it.

I have discussed the "profound question" of cosmic origins with my cosmologist friends. One who believes in the meta-universe idea is Sean Carroll, a physicist at the

California Institute of Technology. Working with Alan Guth, a pioneering cosmologist at the Massachusetts Institute of Technology, Carroll has developed a cosmological theory he calls "Two-Headed Time." In this theory, time has existed forever. So, there was definitely something before the Big Bang. Furthermore, the evolution of the cosmos is symmetric in time, with the behavior of the universe before the Big Bang a nearly mirror image of the behavior after the Big Bang.

In this theory, until 13.8 billion years ago, the universe was contracting. It reached a minimum size at the Big Bang and has been expanding ever since, like a Slinky that falls to the floor, reaches its highest compression upon impact, and then bounces back to larger dimensions. In this model, the universe does not oscillate. There's one contraction and one expansion, coming from infinite time and heading toward infinite time.

Although time has existed forever in the Carroll-Guth model, the *direction* of time has some interesting peculiarities. Theoretical physicists, unlike most other people, spend their time thinking of such things as why we remember the past and not the future. They have concluded that the forward direction of time is determined by the movement of order to disorder. On the whole, the past is more orderly than the present, and the future is less orderly. For example, if you see a movie of a teacup perched on the edge of a table, then falling off, and then shattering on the floor, the movie looks normal; but if you see a movie of the broken shards of a teacup gathering themselves off the floor, jumping up to a table, and assembling a teacup, you would say that movie is being played backward. That's the idea.

Now it's well known in the science of order and disorder that, other things being equal, larger spaces allow for more disorder, essentially because there are more places to lose things. (This can all be made quantitative.) Equivalently, smaller spaces have more order. As a consequence, in the Carroll-Guth picture, the order of the universe was at a *maximum* at the Big Bang, when the universe was most compact, with order decreasing both before and after the Big Bang. *Thus the future points away from the Big Bang in both directions of time.* A person living in the contracting phase of the universe sees the Big Bang in her past, just as we do. When she dies, the universe is larger than when she was born, just as it will be for us. If you think of the Big Bang as a pothole in the infinite road of time, a sign at that location pointing toward the future would point in both (opposite) directions of the road, like the signs pointing to the Emerald City in *The Wizard of Oz*. Science fiction? Perhaps, and perhaps not. Brilliant physicists are contemplating this stuff.

The other possibility is that the universe, and time, did not exist before the Big Bang. In this view, time is not fundamental. Rather it "emerged." Advocates of this hypothesis believe that the universe materialized literally out of nothing, at a tiny but finite size, and expanded thereafter. Such things are possible in quantum physics. But time didn't exist at that time. Just as in Augustine's divine creation of the cosmos, there were no moments prior to the moment of smallest size because there was no "prior." Likewise, there was no "creation" of the universe, since that concept implies action in time. As physicist Stephen Hawking, one of the devotees of this idea, describes it, "the universe was neither created nor destroyed. It would just BE."

One of the first cosmologists to suggest that the universe could appear out of nothing was Alexander Vilenkin, a Ukrainian scientist who came to the United States in 1976 in his mid-twenties and is now professor of physics at Tufts University. Before coming to the United States to do graduate work, Vilenkin saw his acceptance to graduate school in the Soviet Union rescinded, he thinks possibly because he was in disfavor with the KGB. So Vilenkin began work as a night watchman in a zoo—giving him plenty of time to think cosmological thoughts.

Alex Vilenkin is a serious man who, unlike many physicists, does not joke around much, and he takes his work on the universe at t = 0 extremely seriously. "No cause is required to create a universe from quantum tunneling," he once said to me, "but the laws of physics should be there." I asked him what "there" means when time and space don't yet exist, but I didn't get a satisfying reply.

When Vilenkin talks about "quantum tunneling," he is referring to the spooky phenomenon in quantum physics in which objects can perform such magic feats as passing through a mountain and suddenly appearing on the other side, without ever going over the top. That mystifying ability, which has been verified in the laboratory, follows from the fact that subatomic particles behave as if they could be in many places at once. Quantum tunneling is common in the tiny world of the atom but totally negligible in our human world—explaining why the phenomenon seems so absurd. But in the ultra-high-density era of our infant universe, very near t = 0, the *entire universe* was the size of a subatomic particle. Thus, the entire universe could have "suddenly" appeared from wherever things originate in the impossible-to-fathom quantum haze of probabilities.

One of the most detailed models of the universe "during" the quantum era of the Big Bang is due to James Hartle, of the University of California at Santa Barbara, and Hawking, of Cambridge University. Time appears nowhere in Hartle and Hawking's equations. Instead, they compute the probability of certain snapshots of the universe, using quantum physics to do so. One of those snapshots had the highest density of energy. That one could be labeled the "beginning."

Hawking, Carroll, and other quantum cosmologists

are not unaware of the vast philosophical and theological reverberations of their work. As Hawking says in his book *A Brief History of Time,* many people believe that God, while permitting the universe to evolve according to fixed laws of nature, was uniquely responsible for determining those laws. According to this view, God wound up the clock at the Beginning and chose how to set it in motion. Hawking's own theory provides a God-less explanation for how the universe might have wound itself up—by proposing a method of calculating the "early" snapshots of the universe that has no dependence on "initial conditions" or boundaries or anything outside the universe itself. The universe is self-contained. The icy laws of quantum physics are completely sufficient. "What place, then, for a creator?" asks Hawking.

One would expect most "quantum cosmologists" to be atheists, as are the majority of scientists. A prominent exception is Don Page, a leading quantum cosmologist at the University of Alberta and also an evangelical Christian. Page is a master computationalist. When he and I were fellow graduate students in physics at the California Institute of Technology, he used to quietly take out a fine-point pen whenever confronted with a difficult physics problem. Then, without flinching or pausing, he would begin scribbling one equation after another in a dense jungle of mathematics until he arrived at the answer.

Although he has collaborated with Hawking on major papers, Page parts ways with him on the subject of God. As Page once said to me, "As a Christian, I think there is a Being outside the universe that created the universe and caused all things." And in a guest column in Carroll's blog *The Preposterous Universe,* Page sounded like both scientist and theist simultaneously: "One might think that adding the hypothesis that the world (all that exists) includes God would make the theory for the entire world more complex, but it is not obvious that is the case, since it might be that God is even simpler than the universe, so that one would get a simpler explanation starting with God than starting with just the universe."

Most quantum cosmologists do not believe that anything *caused* the creation of the universe. As Vilenkin said to me, quantum physics can produce a universe without cause—just as quantum physics shows how individual electrons can change orbits in atoms without cause. That is, quantum physics can predict how a large collection of electrons and atoms will behave, on average, but not how individual electrons and atoms will behave. There are no definite cause-and-effect relationships in the quantum world, only probabilities. Carroll put it this way: "In everyday life we talk about cause and effect. But there is no reason to apply that thinking to the universe as a whole."

Causality to science is like form to an artist. It's a foundational belief. According to the principle of causality, closely related to the Central Doctrine of Science, each event in the physical world was caused by a prior event (through the action of the laws of nature), and that prior event was caused by a prior event, and so on. In principle, this long, connected chain of causes and effects can be traced back and back through time until one reaches the origin of the universe. At that point, the chain comes to an end. We have reached the edge of the cliff. If one subscribes to the Hartle-Hawking-type picture of a universe with a finite beginning, there is no "before" that point. That point is the earliest and densest configuration of the universe. The universe would just "be." If one subscribes to the Carroll-Guth picture of an eternal universe, one runs into another problem. At its point of maximum compression and density at the Big Bang, the entire universe would likely be subject to the effects of quantum gravity. As I discussed in a previous chapter, at these tiny sizes and enormous densities, gravity and quantum effects combine to cause chaotic fluctuations in time and space. Causality requires an orderly progression of time. When the entire universe is subject to such temporal fluctuations, causality as we know it does not apply.

These issues pose little difficulty to believers in God. The universe was brought into being by God, the first and ultimate cause. But nonbelievers have a great deal of difficulty. It may be that quantum physics can produce a universe from nothing, without cause, but such an accidental and *unanalyzable* origin for EVERY-THING seems deeply unsatisfying, at least to this pilgrim. In the absence of God, we still want causes and reasons. We still need to make sense of this strange cosmos we find ourselves in. Permanent or impermanent, absolute or relative, we still long for answers, and understanding. Evidently, science can find reasons and causes for everything in the physical universe but not for the universe itself. What caused the universe to come into being? Why is there something rather than nothing? We don't know and will almost certainly never know. And so this most profound question, although in tightest embrace with the physical world, will likely remain in the domain of philosophy and religion.

Ants (2)

I think I'll do a careful study of one square inch of Lute Island. Let me gather a few things. Got them. Now I'm off to my favorite mossy spot south of my house, with my lab journal and magnifying glass. In case future scientists want to duplicate my research, the coordinates of my square inch are latitude 43.79982, longitude −69.92186, give or take, and my examination period is beginning at—let me look at my watch—10:45 a.m. Eastern Time, August 22, 2016.

10:45 a.m.: Under my magnifying glass, the moss appears like a micro forest of spruce trees. Most of the micro leaves of each micro branch are green, but some are russet-colored, some saffron. It is a thick little forest, with the branches of neighboring micro trees entwined with each other. Here and there, between the tiny branches, I see tiny bits of seed pods. I wait for some action, but nothing moves. Patience!

10:47 a.m.: My patience has paid off. A tiny russet-colored insect has just appeared. It looks like a cross between an ant and a tick, but far smaller. I estimate its size as about one-fiftieth of an inch. It has multiple legs and antennae. It must be one of the zillions of nameless animals that inhabit every cubic foot of the biosphere. But this is my animal, the first that has appeared in my square inch. I'll designate it Animal A. It moves this way and that, haltingly, trying to find its way through the thicket of micro forest. I wonder where it is trying to get to and where it has come from. It doesn't seem concerned with the Big Bang. After about ten seconds, it disappears, hidden in the micro undergrowth.

10:49 a.m.: A black ant appears, huge under my magnifying glass and gigantic compared to Animal A. While Animal A moved about tentatively and modestly, as if aware of its insignificant size, the ant, which I'll call Animal B, marches ahead with a swagger. Animal B advances as if on a mission. Perhaps this fellow is highly intelligent, a participant in my Smart Ant Conundrum. At this scale of living reality, the ants clearly rule. Animal B, the king, is gone in five seconds.

———

10:50 a.m.: A black insect of some kind suddenly drops down from the sky onto my journal pages. It falls outside of my square inch, my laboratory, so I flick it away without further note and return to my investigations.

10:51 a.m.: Another tiny insect, Animal C, whitish in color, appears deep down in the micro thicket, about ten bug-lengths below the micro treetops. Roughly the size of Animal A, it is also multi-legged, with antennae. Animal C moves much more slowly than the others I've seen. It does not seem to be going anywhere. Ambitionless. Aimless. I've seen the syndrome before.

10:53 a.m.: As the sun moves in and out of the clouds (or the clouds move past the sun, depending on your point of view), my microcosm alternately glows and dims.

10:54 a.m.: For the first time during this observation period, I notice that each branch of each tree in the micro forest is gleaming with dozens of points of light. These points must be tiny shiny surfaces reflecting the sun. For some reason, I didn't see them before.

10:55 a.m.: Another black ant soldiers by, huge and purposeful.

10:56 a.m.: Maybe I'll end these observations. I feel restored. All seems well in the world, at least in one square inch of it.

Multiverse

Let us go back to the meaning of "universe." The word comes from the Latin *unus*, meaning "one," combined with *versus*, which is the past participle of *vertere*, meaning "to turn." Thus the original and literal meaning of "universe" was "everything turned into one."

As astronomers have been able to measure greater and greater distances in space, the contents and size of the universe have grown. Still, we have a conception of a unified whole, a "universe." Most modern physicists and astronomers now define the universe as the totality of time and space that can communicate with itself at any time from the infinite past to the infinite future. Furthermore, it is assumed that the laws of nature and other fundamental parameters, such as the speed of light, are everywhere the same in the "universe"; this is part of the Central Doctrine of Science. In that sense, the universe is indeed a Unity.

Recent developments in science have suggested the possibility of a multiplicity of universes, called the "multiverse." First, new theories of physics—one called "eternal inflation" and another called "string theory"—have *predicted* other universes in addition to our own. While these theories are speculative and unproven, many physicists worldwide have devoted their careers to them.

Eternal inflation is a variant of the "inflation" cosmological model first proposed in the early 1980s. That model, a modification of the standard Big Bang model, has successfully met a number of experimental tests and is assumed true by most physicists today. However, the original inflation model concerns only our universe. The eternal inflation model, an extension of the original inflation model, predicts that the same physics that propels the exponentially rapid expansion of an inflating universe can spawn new universes, which branch off from the original.

String theory was originally proposed in the 1970s as a theory of the strong nuclear force, then enlarged to become a theory of all the forces of nature. In string theory, there are seven extra dimensions of space in addition to the three that we are familiar with. We don't observe these extra dimensions, according to the theory, because they are curled up into ultra-tiny loops. It turns out that there are zillions of different ways that the extra dimensions can be folded up, and each way corresponds to a different universe, with different properties. So far, string theory has not managed to predict anything that could be tested. Potentially, it could make some predictions about *our* universe, such as that it must have certain symmetries, and those predictions, if confirmed, would give us some confidence in the theory. At the present, we must regard the theory as speculative. In any case, we will almost certainly never be able to confirm or refute the existence of other universes, even though they may be predicted.

But beyond these predictions is another argument for the multiverse. The existence of other universes has been invoked as a plausible explanation for the observation that our universe seems to be "fine-tuned" to allow the existence of life. That is, if various fundamental and fixed parameters of the cosmos, such as the strength of the nuclear force or the density of cosmic "dark energy," were a little larger or a little smaller than they actually are, life could not have arisen in our universe—not just life similar to that on earth but life of any kind. For example, consider the so-called dark energy, an exotic form of energy that, unlike other forms of matter and energy, leads to a *repulsive* gravitational force. If the density of dark energy—which has been measured to have a value of about 6×10^{-9} ergs of energy per cubic centimeter of space—were a little larger, then the universe would have expanded so rapidly that matter could never have coalesced into stars. On the other hand, if the density of this cosmic energy were a little smaller (and negative), then the universe would have recollapsed long before stars could form. In other words, stars cannot form unless the density of dark energy lies within a relatively narrow range. Although we are not certain what conditions are needed for life, we are almost certain that atoms are needed, and we have strong evidence that all atoms heavier than hydrogen and helium were made in the nuclear furnaces of stars. Without stars, no atoms. Without atoms, no life. And without a carefully tuned density of dark energy, no stars.

The question, of course, is: Why? Why should the universe care about life? The multiverse solves this conundrum. If there are lots of other universes with different properties—some with values of dark energy larger than in our universe and some with smaller—then some universes have stars and life, and some do not. By definition, we live in one of the universes that permits life—because if we didn't, we wouldn't be here to discuss the matter. In an analogous manner, one can ask why it is that we live on a planet that is the right distance away from Sol to have liquid water. If we were a bit closer, all the water would evaporate in the high heat, and if we were a bit farther, it would freeze in the cold. The ancient Roman physician and philosopher Galen and many others argued that the fortuitous position of the earth is due to the "benevolent influence" of the gods. But a more satisfying answer (to most scientists) is simply that there are lots of planets in the galaxy. By the odds, some fraction of them should happen to be the right distance from their central star to allow liquid water. We live on such a planet, because if we didn't . . .

The multiverse idea is fascinating for several reasons. For one, it represents a monumental extension of our conception of existence and reality. For another, it is an idea that almost certainly cannot be confirmed. By definition, different universes cannot contact each other. Thus, it is hard to see how we could ever experimentally determine that other universes exist. One of the notions that distinguishes science from religion and philosophy is that all of the beliefs of science must be testable by experiment in the external world. If the multiverse idea cannot be tested, is it scientific?

Another important issue is that if the multiverse idea is true, then our universe is simply an accident, one universe among many possible universes, a roll of the dice. The historic mission of physics has been to explain every fundamental aspect of the physical universe as a *necessary* consequence of a few fundamental laws and principles, like a crossword puzzle with only one solution. But in the multiverse, the same fundamental laws lead to many different universes. We must accept certain aspects of our universe, such as the value of the dark energy, as accidents. For all of these reasons, the multiverse has divided the community of physicists. Some reluctantly support the idea as the most plausible explanation of the fine-tuning problem. Others reject it as nonscience or extreme speculation.

Lastly, the multiverse idea is a sharp blow against the ideal of Unity in the grandest possible terms. If the multiverse exists, our universe is only one of a multitude of universes. Unity surrenders to multiplicity.

Paradoxically, the ideal of Permanence is restored. Because even though everything in our universe must pass away (even in the "Two-Headed Time" theory of Carroll and Guth), the uncountable number of universes in the multiverse, as a group, would be eternal. New universes have been born forever in the past, are being born now, and will forever be born in the future. Likewise new planets and stars. New oceans. New life. Whatever space-time continuum the multiverse inhabits would exist forever, from the infinite past to the infinite future. To borrow some earlier language, that continuum would be the womb of all universes. That continuum would be as close as science can come to Anaximander's ageless infinite, or to the One of the Hindu Rig Veda, or to the "formless void" of Genesis.

Previously, I have said that science endorses very few of the Absolutes. The Central Doctrine of Science is one. A "final theory" is another. But as I have further examined the body of science in these pages, I must slightly revise my earlier statements. (And in the act of revision, am I not acting like a scientist?) Experimental and theoretical investigations destroyed the iconic idea of the constancy of stars but, in the process, renewed the idea of the constancy of energy. Considerations of the multiverse destroy the idea of the unity of a universe, but introduce the idea of permanence in the totality of universes. It would seem that even though we live in a material and fleeting world, the Absolutes are inescapable. The Absolutes keep circling through our theories. The Absolutes are deeply lodged in our imaginations, our longings and hopes, our way of comprehending existence.

Humans

One of the most arresting images of the civil rights movement is Ernest Withers's photograph of the black sanitation workers' strike in Memphis. The photo was taken on March 28, 1968, a week before Martin Luther King Jr. was assassinated in that roiling city. In previous months, the city had rejected the requests of the black sanitation union—requests, for example, that black garbage workers be paid the same wages as white garbage workers. The photograph shows a few hundred black men gathered on the street in front of the Clayborn Temple for a solidarity march. Many are wearing nice jackets and pants. The men do not appear angry. But they do look totally committed to their cause, and they look proud. With quiet grace, each of them is holding an identical sign, which has just four words: I AM A MAN. The four words are repeated a hundred times in the photograph, on a hundred white placards, and the silence is deafening.

I grew up in Memphis. When I look at this photograph, I am reminded that a desire for dignity is part of human nature. I am also provoked to think about what it means to be a human, period, of any color or creed. Is there anything unique about humanity, distinguishing us from other living creatures on earth? Do human men and women occupy a special place in the cosmos, as described in the Bible and the Qur'an? And what of the future? Where are we headed, we human beings?

Lute Island. Our children grew up spending summers at Lute Island. They have the island in their blood. They know the speckled dark of the night skies and the rhythm of tides and the silences. Now, our children have children. Our granddaughter Ada spent her first summer at Lute Island at the age of two months. Now, at age three years, she considers the island her second home. In just one generation, Ada's world is vastly different from the world of her mother's childhood. When Ada is at her winter home, in New York, she frequently FaceTimes with me and is quite accustomed to talking to my tiny image on her mother's iPhone. What would have been a sorcerer's magic forty years ago is commonplace for her. Actually, I can't imagine what's in her head because she has grown up with the internet and advanced communication technology. She tells me stories on FaceTime, laughs when I make funny faces, and carries me around with her as she walks from room to room or outside in the park. Having a miniature, two-dimensional version of me in her pocket is part of her unquestioned reality. For her, and others in her generation, the separations of space and time have mostly dissolved. For her, I can be in two places at once. For her, the fleshy version of me is only Grandfather 1.0.

If I ask what it means to be human in the year 2017 and where we are headed, Ada's techno-reality must be part of the answer. Right under our noses, *Homo sapiens* is transitioning into *Homo techno*. And the change is happening not over millions of years. It's happening in single human lifetimes, by our own inventions and technology. We are modifying our evolution by our own hand. We are remaking ourselves. Nothing could challenge the permanence and constancies of the Absolutes more than our own evolution and change.

One of the first people to foresee the power of technology to change human identity was the early-seventeenth-century English philosopher and scientist Francis Bacon. I have long been a fan of Mr. Bacon. Despite a controversial career as a member of Parliament, dogged by accusations of corruption and public impropriety, Bacon is widely credited with introducing the modern scientific method. That notion says that we should examine statements of purported fact with skepticism and believe only what we can verify with our own observation.

Let me reach behind my desk to Dr. Eliot's "five-foot shelf of books," as I've done many times before, and pull out volume 3, containing Bacon's utopian novel, *The New Atlantis*. Here, well before the scientific revolution, Bacon imagines a college of the future, called Solomon's House, in which new inventions vastly expand the physical capabilities of human beings:

> We procure means of seeing objects afar off; as in the heavens and remote places . . . We have also helps for the sight, far above spectacles and glasses in use. We have also glasses and means to see small and minute bodies perfectly and distinctly; as the shapes and colours of small flies and worms, grains and flaws in gems, which cannot otherwise be seen. . . . We have also sound-houses, where we practise and demonstrate all sounds, and their generation . . . We have certain helps which set to the ear do further the hearing greatly.

I wonder what Bacon would say about today's scientific instruments that can record X-rays and radio waves, or microscopes able to see single atoms, or telescopes able to read the writing on a dollar bill several miles away, or microphones able to hear the footsteps of an ant. Far beyond Bacon's imagination would have been the computers of today, or our ability to implant a silicon chip into the brain of a paralyzed person that can detect his desires and then activate a robot. Certainly, Bacon could not have conceived of FaceTime.

As I gaze out at the quiet path leading from the house to the ocean, let me speculate on the human beings of the future, my own Solomon's House. With the galloping pace of invention, I dare not look further than a hundred years ahead. In a century or less, we may have special lenses implanted in our eyes that do what our external detectors do now, allowing us to see X-rays and other frequencies of light much higher than the visible part of the electromagnetic spectrum. With this technology and off-the-shelf X-ray emitters, we will be able to see through clothing, walls, and many other surfaces that ordinary light cannot penetrate. A hundred years from now, we may have computer chips implanted in our brains that connect our minds directly to the internet. In such a situation, we may need only think of a piece of information we're seeking and it will be instantly transmitted to our brains. Using the same technology, we might be able to communicate directly with the minds of other people through the internet. Such a scenario would raise vast new issues of privacy and intellectual property, and probably require new kinds of laws and legal constructions. Fifty years from now or less, we may have tiny robots, the size of red blood cells or smaller, that can be injected into our bodies to kill cancer cells, deliver drugs with highly targeted precision, repair damaged or faulty DNA, and greatly enhance our immune system. With developments in neuroscience and the understanding of memory storage, we may have brain implants that teach us new languages in minutes. Words in Chinese or Swahili could be streamed to our brains in real time. And if we wish to utter those words, instructions for the unfamiliar movements of mouth muscles could also be streamed in real time.

With an understanding of how information is embodied in neurons and synapses, we may someday be able to digitize a large fraction of the memories in a human brain and have that information stored on an external computer. In such a situation, we could re-create much of "ourselves" outside our physical bodies. Quantum physics and the Second Law of Thermodynamics would prevent such re-creation from being total, so that we would not be able to tele-transport ourselves whole as do Captain Kirk and his crew on *Star Trek*. But large parts of our memory and experience might be preserved and transported, so that there could be many "versions" of us contained in computers. Conversely, memories of events that never happened could be downloaded into our brains, so that we could have the sensation of any of a large number of past lives and identities.

These are just a few ideas for my Solomon's House, none of them prevented by scientific limitations as far as I know. In short, the human being of the future, *Homo techno,* will be part animate and part inanimate, a hybrid of living animal and machine, a heart and soul fused to a computer chip. Everything will be changed. Everything is already changing. As J. B. S. Haldane, one of the first prophets of so-called "transhumanism," said nearly a century ago, "Science is as yet in its infancy, and we can foretell little of the future save that the thing that has not been is the thing that shall be; that no beliefs, no values, no institutions are safe." Everything will be changed.

Many people in the United States, and in the world as a whole, either deny that human beings are evolving or are uncomfortable with that fact. A recent Gallup survey found that 42 percent of Americans believe that humans were created in their present form at the same time the world was created. I would argue that such a belief is part of a commitment to the Absolutes, a personal and visceral version of constancy, permanence, and certainty. Although the kind of technological evolution I have described above is not the same as the transition from amoebae to fish to *Homo sapiens,* when technology is used to alter our bodies and brains, I believe that development should be considered as part of our evolution.

It is an interesting question why so many of us resist

the idea of the evolution of human beings. There are psychological, theological, and even biological reasons. The biological is perhaps the most basic. One of the definitions of a living organism, at a primitive level, is the ability to separate itself from its surroundings and to create a stable and orderly environment. A single cell, or a mammal, cannot survive if it is constantly subject to shocks and disturbances. We could not survive if our DNA were constantly changing, or if our cell walls were constantly dissolving. The need for some kind of equilibrium, order, homeostasis, within ourselves and our immediate environment must be buried deep in our psyche. Yet in the rapidly changing interface between humans and technology today, we are far from equilibrium.

The psychological. Aside from theological considerations and despite all that modern science has done to show that *Homo sapiens* are animals, I would argue that we still entertain a certain amount of what might be called "species chauvinism." If you Google any phrase involving "animal," you will be led to a website that involves nonhuman animals but not human animals. And just consider how we treat many nonhuman animals. We believe that we are a superior and special species, that we have an elevated place among living creatures on this planet. "God made man in his own image . . . And God said to [Adam and Eve] 'Be fruitful and multiply and . . . have dominion over the fish of

the sea and over the birds of the heavens and over every
living thing that moves on the earth.'" At some deep
level, that elevated position is part of our self-identity,
our relationship to other living things on the planet.

That self-identity, in turn, is challenged by the idea
that *Homo sapiens* is a passing phase in the story of evolution on earth. If we view the evolution of life as a
long line of falling dominoes, with the dominoes that
have already fallen representing the past, the dominoes
still standing the future, and the domino just beginning
to fall the present era, then *Homo sapiens* is that domino
just starting to fall. Our moment in history is passing as
we speak. Ahead, and not far ahead, is *Homo techno.* We
have hints of *Homo techno,* but we don't know what he/
she/it will be. One thing we do know, as Haldane said,
is that it won't be us. Perhaps parts of us. To many of
us, including myself, this picture disturbs. Why it disturbs me I can't exactly say. Why should I care if *Homo
sapiens* doesn't exist a few centuries from now? But
I do.

Finally, there is the theological basis for rejecting
human evolution, in some ways the easiest to fathom.
The sacred books in Christianity, Judaism, and Islam
share a common Creation story, and all declare that the
first human beings were created by God in their present form, soon after the Beginning. For example, here
are the words in the Qu'ran:

O mankind! reverence
Your Guardian-Lord,
Who created you
From a single Person [Adam]
Created, of like nature,
His mate [Eve], and from them twain
Scattered (like seeds)
Countless men and women

Hinduism does not have a single creation story, nor does it subscribe to a single god. However, Brahma, the Creator, sits at the top of the pyramid of Hindu divinities and, according to the Matsya Purana, Brahma created the goddess Saraswati, mated with her, and thereby produced the first man, Manu. Later came the first woman, Ananti. From Manu and Ananti descended the rest of us.

For those who take the Bible or the Qur'an or the Matsya Purana as the true word of an omniscient God, the case is closed. We have learned from God Himself that *Homo sapiens* has not evolved up to this point in the history of the world. Given that belief, the idea that *Homo sapiens* may evolve in the future into *Homo techno,* or anything else, is all the more difficult to embrace. And here we have a direct confrontation with science. How is that conflict resolved? For those who take the sacred books as the word of God, the resolution is simple. Divine action and miracles lie outside of science. Writes the contemporary Islamic scholar Sheikh Abdul Wahhab al-Turayri in *Islam Today:*

> The direct creation of Adam, peace be upon him, can neither be confirmed nor denied by science in any way. This is because the creation of Adam was a unique and singular historical event. It is a matter of the Unseen and something that science does not have the power to confirm or deny. As a matter of the Unseen, we believe it because Allah informs us about it. We say the same for the miracles mentioned in the Quran. Miraculous events, by their very nature, do not conform to scientific laws and their occurrence can neither be confirmed nor denied by science.

I have difficulty with Sheikh Abdul Wahhab al-Turayri's statement for the following reason. I can grant him that the origin of the universe might have been a one-time, singular cosmic event, unknowable by science. Science theorizes about what created the universe, but science cannot test those theories with any certainty. (See the earlier chapter "Origins.") Still, humans, monkeys, reptiles, and fish live in the physical world. And there is a huge amount of physical evidence—from fossils to embryonic development to

comparative DNA analyses, to analyses of the changing chemical composition of the planet—pointing strongly to an evolution of living forms over time, starting with primitive one-celled organisms that could survive in an atmosphere without oxygen and ending with human beings. The denial of evolution amounts to the statement that all of the many pieces of longitudinal evidence from biology and physics and chemistry that fit together are a coincidence. Not one coincidence, like a one-time creation event, but thousands of coincidences over the ages, equivalent to thousands of miracles extending over a long span of time.

In my pastoral descriptions of Lute Island, I forgot to mention one detail: telephone poles. When the six families who own the island had our first meeting more than twenty-five years ago, we voted to bring electricity to the island. The vote was not unanimous, but it carried. As you walk down the central path of the island, in addition to seeing trees and moss you see telephone poles. Aware that the poles are not indigenous to the island, we planted them at irregular intervals and in a randomly curving line, to give a more "natural" appearance. Of course, our houses, made of concrete, machined lumber, drywall, and glass, are also not indigenous to the island.

All of this discussion raises the question of what we

call "natural" and what we call "unnatural." This question is of vital importance as we make our way from *Homo sapiens* to *Homo techno*. As we modify our bodies and brains with technology, are we creating unnatural abominations of ourselves? Are we deviating from God's image?

I still remember the public outcry when the first mammal was cloned in 1996, a sheep named Dolly, who was created by taking the nucleus of a cell of an adult sheep, transferring it to a sheep egg cell with no nucleus, and giving the freshly nucleated egg an electric shock, as did Dr. Frankenstein. Around the world, some people thought that human beings were transgressing into territory that was none of our business, that we were not entitled to create life, that the biologists at the Roslin Institute in Edinburgh had committed an *unnatural* act.

What does "unnatural" really mean? Does it strictly mean "against nature"? Does it mean a condition not intended by God? Does it mean an activity that somehow falls outside of the innate biology of *Homo sapiens*? Or possibly all of the above?

I suggest that with the advances of science and technology, our species has already gone so far beyond many of the physical limitations associated with our innate biology that the distinction between "natural" and "unnatural" has begun to dissolve. One might begin with the simple feature of our life span. In the year 1800,

the life expectancy of Americans, averaging males and females, was about thirty-seven years; by 1900 that number had increased to forty-seven years. Principal causes of death over the centuries have been smallpox, influenza, tuberculosis, pneumonia, malaria, typhoid, and gastrointestinal infections. With the understanding of the germ theory of disease, innovations in public health, and the discovery of antibiotics and vaccines in the twentieth century, the life expectancy of Americans is now seventy-nine and rising.

Then the technology. The hearing aids and eye-glasses that Francis Bacon imagined are now a reality. Should we consider such technology natural or unnatural? If you object to the question by claiming that eyeglasses only restore what is natural, consider microscopes, which allow us to see things thousands of times smaller than could be seen with the unaided eye. And X-ray and infrared cameras, which allow us to detect wavelengths of light far beyond the "visible" sensitivity of the rod and cone cells in our eyes. Or airplanes, which allow us to fly about like birds. Without a doubt, we live in a manufactured world—not only the microscopes and sonar that extend our senses, but our cities, our cars, our temperature-controlled environments, our iPhones. It's all made by us.

Either all of it is "unnatural," or all of it is "natural." If a giant intelligent being were looking at earth from deep space, it would see a throbbing dance of activity and construction: ants building elaborate tunnels underground, beavers building mud dams on ponds, bees building honeycombs with beautiful hexagonal cells. And human beings building houses and cities. All of us living creatures are modifying our environments. True, the giant being would see an acceleration of activity in the last couple of centuries. Here's the way that I think of it: Since all of our inventions come from our brains, and since our brains evolved from "natural" processes, then everything we invent, including modifications to our own bodies, should be considered "natural." In this sense, *Homo techno* is no more unnatural than *Homo sapiens* today.

"Maybe," I hear you say grudgingly. "But then what does it mean to be *human*? Isn't there some group of qualities, some desires or inclinations or values, some emotional or spiritual core, that will still define us as *human* after we have tiny robots in our bloodstream and silicon chips in our brains?" Good question.

I began with Withers's photograph, which suggests that dignity is part of humanness. In H. G. Wells's iconic novel *The Island of Dr. Moreau,* the formerly celebrated physiologist Dr. Moreau performs outlawed vivisection experiments, creating beings half human and half non-human. For example, parts of a bear might be grafted onto parts of a man. A colony of these hybrid beasts lives in the jungle. Their leader is a large gray thing named Sayer of the Law, who recites a strange chant called the "Law." The law prohibits "bestial" behavior:

> Not to go on all-fours; that is the Law. Are we not Men? Not to suck up Drink; that is the Law. Are we not Men? Not to eat Fish or Flesh; that is the Law. Are we not Men? Not to claw the Bark of Trees; that is the Law. Are we not Men? Not to chase other Men; that is the Law. Are we not Men?

What does "Are we not Men?" mean here? Precisely that these creatures are not nonhuman animals. The law consists completely of negatives, things the creatures are not supposed to do, which are the things that nonhuman animals do. Wells is defining humans by what they are not.

I will approach this difficult question with one more example. In the film *Star Trek II: The Wrath of Khan*, James Kirk (who has now been promoted to admiral) comes aboard the *Enterprise* spaceship to inspect a training simulation. There he meets Spock (now captain of the *Enterprise*) and several trainees, including Saavik, who is half Vulcan and half Romulan. As nearly everyone knows who has watched television in the last fifty years, Spock himself is half Vulcan and half human and is always struggling with the conflict between the totally logical and emotionless Vulcan side of him and the emotional human side of him. Here's the scene:

SPOCK: Open the Air Lock.

KIRK: Permission to come aboard, Captain?

SPOCK: Welcome, Admiral. I think you know my trainee crew. Certainly they have come to know you.

KIRK: Yes, we've been through death and life together. Mister Scott, you old space dog. You're well?

SCOTT: I had me a wee bout, sir, but Doctor McCoy pulled me through.

KIRK: Oh? A wee bout of what?

McCOY: Shore leave, Admiral.

KIRK: Ah. And who do we have here?

PRESTON: Midshipman First Class Peter Preston, engineers mate, sir!

KIRK: First training voyage, Mister Preston?

PRESTON: Yes sir!

KIRK: Well, . . . shall we start with the Engine Room?

SCOTT: We'll see you there, sir, and everything is in order.

KIRK: That'll be a pleasant surprise, Mister Scott.

SPOCK: I'll see you on the bridge, Admiral. Company dismissed!

SAAVIK: (in Vulcan) He's never what I expect, sir.

SPOCK: (in Vulcan) What surprises you, Lieu-
   tenant?
SAAVIK: (in Vulcan) He's so . . . human.
SPOCK: (in Vulcan) Nobody's perfect, Saavik.

Here we have the writers of *Star Trek* opining on
what it means to be human. And what's so "human"
about Kirk? He's warm, humorous, playful, expressive. He jokes around with Mister Scott. So, are these
qualities part of what defines us as humans? Are these
qualities uniquely human? Not by the evidence. Anyone who owns a dog knows that dogs can be loving,
playful, happy, sad, expressive, and all the other qualities that Saavik was referring to. (Perhaps there were
no dogs on Saavik's home planet.)

In his book *The Descent of Man,* Darwin noted the
similarity between man and other animals in their
emotional capacities: "the lower animals, like man,
manifestly feel pleasure and pain, happiness and misery. Happiness is never better exhibited than by young
animals, such as puppies, kittens, lambs, etc., when
playing together, like our own children." A great many
experiments have strongly confirmed these similarities.
For example, more than fifty years ago, researchers at
Northwestern University Medical School found that
macaques refused to pull a chain that delivered food
to themselves if doing so also caused a companion
to receive an electric shock. And as far as raw intelli-

gence goes, numerous experiments and observations of monkeys, ravens and crows, dolphins, whales, and other nonhuman animals show without a doubt that these creatures have the ability to solve problems, to communicate, to play, to recognize themselves in the mirror, and to do other activities we associate with intelligence.

Finally, let me come back to the Withers photograph and dignity. Dignity is a fine and subtle quality that does not quite fit into the categories of emotional capacity or intelligence. But dignity must surely be related to a sense of self, and many nonhuman animals have self-awareness. (Consider the dolphins, who recognize themselves in a mirror.) I don't know for sure whether dolphins and dogs and crows have or desire dignity, but I do know that many of the (nonhuman) animal protection organizations use that particular word to describe how we should treat other animals.

The uniqueness and specialness of *Homo sapiens* may be the final Absolute. But considering all I have pondered these long summer days, I must reluctantly conclude that there seems to be nothing unique about us human beings. And thus nothing unique to pass on to *Homo techno*. The dominoes are falling, and our moment is passing. Surely, there will remain in our evolutionary descendants some capacity for sadness and joy, empathy, love, guilt and anger, playfulness, creative expression. But those qualities will not necessarily point back to us. Instead, they will point back to a large family of noisy and feeling animals—the living, throbbing kingdom of life on our planet, of which we are a part. A kingdom that has never stood still but has changed and evolved and will always do so. A kingdom that consecrates life and its possibilities even as each of its individuals passes away. A kingdom that dreams of unity and permanence even as the world fractures and fades. A kingdom redesigning itself, as we humans now do. All is in flux and has always been so. This is the thing that I know, and perhaps the only thing that I know. Flux is beyond sadness and joy. Flux and imper-manence and uncertainty seem to be simply what is. At least in the physical world.

I will end this day listening to Bach's exquisite Mass in B Minor. Written to celebrate the Christian God, I will take it to celebrate all gods, for the gods of our faiths are not so different from each other. I will take it to cel-ebrate those who believe and those who do not, for we all want to believe something. I will take it to celebrate life in its myriad forms, even as that life passes away. I will take it to celebrate meaning, even if that meaning is only the moment. The moment is now. As I gaze out the window, a slender blue heron lifts off from the shore and glides over the bay.

Acknowledgments

I am pleased to acknowledge many people who helped me with this book. For conversations on spiritual matters, I thank Rabbi Micah Greenstein, the Venerable Yos Hut Khemacaro, Rabbi Sandy Sasso, Rabbi Dennis Sasso, and Owen Gingerich. For scientific conversations, I thank Don Page, Sean Carroll, Alex Vilenkin, Alan Guth, and Robert Jaffe. For matters in the history of science, I thank Owen Gingerich. For philosophical conversations, I thank Ned Hall and Jeff Wieand. For conversations on psychology and mind, I thank Nick Browning and Ross Peterson. Many others have shaped me in ways big and small.

I thank my literary agents and counselors Jane Gelfman and Deborah Schneider for their steady and loving trust in me over so many years. I thank my longtime and wonderful editor at Pantheon, Dan Frank, who has the rare courage to follow his own judgment instead of the market, who says what he likes and what he doesn't, and who has often pushed me to do better.

Finally, I thank my dear wife, Jean, my loving companion through this fleeting life, and my daughters, Kara and Elyse, who have made me proud and give me hope for the future.

Notes

Longing for Absolutes in a Relative World

9 Plato discussed absolute justice: See *The Republic*, Book V, section 472.

9 Saint Augustine ascribed absolute truth: See *To Consentius, Against Lying* (*Contra mendacio*), paragraph 36.

10 "in its own nature": Isaac Newton, *Principia* (1686), trans. Andrew Motte (Berkeley: University of California Press, 1934), p. 6.

10 "The loveliest music": C. P. Cavafy, "Impossible Things" (1897), in *Complete Poems*, trans. Daniel Mendelsohn (New York: Knopf, 2009), p. 294.

10 "And having made [the universe]": Plato, *Timaeus*, trans. Benjamin Jowett, Great Books of the Western World, vol. 7 (Chicago: Encyclopaedia Britannica, 1952), p. 452.

11 A new survey of 35,000 adults: "America's Changing Religious Landscape," May 12, 2015, at http://www.pewforum.org/2015/05/12/americas-changing-religious-landscape/.

11 A somewhat older survey by the Barna Group: "Religious Beliefs of U.S. Adults (1997 to 2014): Does Absolute Moral Truth Exist?," at http://www.religioustolerance.org/chr_poll5.htm.

11 A 2014 Gallup Survey: "In U.S., 42% Believe Creationist

View of Human Origins," June 2, 2014, at http://www
.gallup.com/poll/170822/believe-creationist-view-human
-origins.aspx.

15 a recent study by Rice University sociologist: Elaine Howard Ecklund, *Science vs. Religion: What Scientists Really Think*
(Oxford: Oxford University Press, 2010).

Material

20 "Nature," wrote Emily Dickinson: Emily Dickinson, poem
668, in *The Complete Poems of Emily Dickinson,* ed. Thomas H.
Johnson (Boston: Little, Brown, 1960).

21 "from so simple a beginning": Charles Darwin, *Origin of
Species* (1859), in Great Books of the Western World, vol. 49
(Chicago: Encyclopaedia Britannica, 1952), p. 243.

23 "Verily, verily": John 3:5, King James Version (hereafter KJV).

24 "In living nature the elements": Jöns Jacob Berzelius, *Lärbok i kemien,* trans. and quoted in Henry M. Leicester, "Berzelius," *Dictionary of Scientific Biography,* vol. 2 (New York:
Scribner's, 1981), p. 96a.

25 "I can't believe that we are just flesh": AL interview with
Rabbi Micah Greenstein, January 5, 2016.

25 "capable of such lofty and astonishing things": Marilynne
Robinson, "Can Science Solve Life's Mysteries?," *The
Guardian,* June 4, 2010, https://www.theguardian.com
/books/2010/jun/05/marilynne robinson science religion.
This essay is extracted from Robinson's book *The Absence of
Mind* (New Haven, CT: Yale University Press, 2010).

26 "It feels a little heavier": Robert Tools, "Heart Recipient
Has Whirr in Chest," Reuters News Service, August 22,
2001, http://www.deseretnews.com/article/859859/Heart
-recipient-has-whirr-in-chest.html?pg=all.

26 the brain of Erik Sorto: ABC News, "Paralyzed Man Drinks

Beer by Moving Robotic Arm with His Mind," May 21, 2015, http://abcnews.go.com/Health/paralyzed-man-drinks -beer-moving-robotic-arm-mind/story?id=31214663; original news story from Caltech: https://www.caltech.edu /news/controlling-robotic-arm-patients-intentions-46786.

29 Peary wrote in his journal: "Diary of Robert E. Peary," at https://catalog.archives.gov/id/304960; "Robert E. Peary's Journal," at http://www.bowdoin.edu/arctic-museum /northward-journal/robert-peary.shtml.

Hummingbird

32 "air within the air": Pablo Neruda, "Ode to the Hummingbird." The poem can be found on Hummingbird-guide.com, at http://www.hummingbird-guide.com/pablo -neruda-hummingbird-poem.html.

33 the required heart rate: The aerodynamic lift pressure is roughly ρv^2, where ρ is the density of air and v is the velocity of air over the wing (proportional to the wing velocity). The exact amount of lift pressure depends on the shape of the wing. This pressure multiplied by the wing area is the upward lift force, which must counterbalance the hummingbird's weight, a few grams. Equating the two gives the required minimum wing velocity and hence flapping rate, assuming a circular motion of the wing.

Stars

34 "Does not the heaven arch itself": Goethe, *Faust,* trans. A. Hayward (New York: D. Appleton and Company, 1840), p. 141.

34 *Sidereus Nuncius* in its original Latin: Galileo Galilei, *Sidereus*

Nuncius, or The Sidereal Messenger (1610), trans. and with notes by Albert Van Helden (Chicago: University of Chicago Press, 1989). I am indebted to Albert Van Helden for his excellent commentary in this edition.

37 "ethereal quintessence of Heaven": John Milton, *Paradise Lost* (1667), Book III, Harvard Classics, vol. 4 (New York: P. F. Collier & Son, 1909), p. 153.

38 "fearefull abounding at this time": James, *Daemonologie*, Project Gutenberg Literary Archive Foundation, June 29, 2008, www.gutenberg.org/catalog/world/readfile?fk_files =845529. See also Geoffrey Scarre and John Callow, *Witchcraft and Magic in Sixteenth- and Seventeenth-Century Europe* (Hampshire, UK: Palgrave, 2001).

38 "I do not wish to approve": Letter from Paolo Gualdo to Galileo, *Le Opere di Galileo Galilei*, National Edition, ed. Antonio Fawaro, 20 vols. (Florence: G. Barbera, 1929–39), 2:564.

39 "My dear Kepler": Galileo, *Opere*, 10:423.

39 "SIDEREAL MESSENGER, unfolding": *Sidereus Nuncius*, p. 26.

40 "Anyone will then understand": Ibid., p. 36.

40 "the highest bell towers of Venice": Galileo, *Opere*, 10:253. See also *Sidereus Nuncius*, trans. and ed. Van Helden, p. 7.

42 "I know that your Reverences": Galileo, *Opere*, 11:87–88. See also *Sidereus Nuncius*, trans. and ed. Van Helden, p. 110.

43 There were many other such fantasies: An excellent discussion of many of the fantasies about the moon and planets after Galileo's findings can be found in Marjorie Hope Nicolson, *Voyages to the Moon* (New York: Macmillan, 1960).

44 "there can be an infinite number": Giordano Bruno, *On the Infinite Universe and Worlds* (1584), trans. Scott Gosnell (Port Townsend, WA: Huginn, Munnin & Co., 2014), Second Dialogue, p. 76.

49 "For surely a dread holds all mortals": Lucretius, *De Rerum Natura* (ca. 60 BC), Book 1, vv. 146–58, trans. and ed. W. H. D. Rouse and M. F. Smith (Cambridge, MA: Harvard University Press, 1982), pp. 15–17.

Atoms

54 "It seems probable to me": Isaac Newton, *Optics,* Book III, Part 1, trans. Andrew Motte and rev. Florian Cajori, in Encyclopaedia Britannica Great Books of the Western World, vol. 34 (Chicago: University of Chicago Press, 1987), p. 541.

55 Pleasing substances are made: Paraphrase of Lucretius, *De Rerum Natura,* Book 2, vv. 398–407. See, for example, Lucretius, *De Rerum Natura,* trans. and ed. W. H. D. Rouse and M. F. Smith (Cambridge, MA: Harvard University Press, 1982), p. 127.

56 At an engaging internet site: "J. J. Thomson Talks About the Size of the Electron," at http://history.aip.org/history/exhibits/electron/jjsound.htm.

58 Becquerel believed that the mysterious radiation: See, for example, American Physical Society News, https://www.aps.org/publications/apsnews/200803/physicshistory.cfm.

59 "As history unveiled itself": Henry Adams, "The Grammar of Science," in *The Education of Henry Adams* (1903; Boston: Houghton Mifflin, 1918), p. 458.

60 "It was quite the most incredible event": Ernest Rutherford in *Background to Modern Science,* ed. Joseph Needham and Walter Pagel (Cambridge, UK: Cambridge University Press, 1938), p. 68.

63 "I could be surprised": AL interview with Jerry Friedman, May 28, 2004.

64 "The explanation of everything is sought": Leo Tolstoy,

My Religion, On Life, Thoughts on God, On the Meaning of Life,
trans. Leo Weiner (New York: Colonial Press, 1904), p. 402.

66   space is not continuous: The most developed theory is called
quantum loop gravity. See Lee Smolin, "Atoms of Space and
Time," *Scientific American,* January 2004.

67   Researchers using NASA's Fermi Gamma-ray Space Telescope: See Vlasios Vasileiou, Jonathan Granot, Tsvi Piran,
and Giovanni Amelino, "A Planck-Scale Limit on Spacetime
Fuzziness and Stochastic Lorentz Invariance Violation,"
Nature Physics 11 (2015): 344–46.

Ants

69   "one feels tense, annoyed, and dissatisfied": Irvin Yalom,
Existential Psychotherapy (New York: Basic Books, 1980), pp.
462–63.

73   Maybe the moment is all: Recent research has shown that
human beings constantly think about the future in a way
rarely found in other animals. See Martin E. P. Seligman,
Peter Railton, Roy F. Baumeister, and Chandra Sripada,
Homo Prospectus (Oxford: Oxford University Press, 2016).
Evidently, constant cogitation about the future either was
developed for survival benefit in our evolutionary history
or is an inevitable by-product of an intelligent brain. In
either case, in my desire to live in the moment, I am fighting
against the natural propensity of the human brain.

Monk

75   "Buddhism is in complete agreement with science": This
and other comments by Yos Hut Khemacaro from interview by AL on January 5, 2016, Phnom Penh.

Truth

78 "If a man proceeded by force": O. R. Gurney and S. N. Kramer, "Two Fragments of Sumerian Laws," *Assyriological Studies*, no. 16 (April 21, 1965): 13–19.

78 "Thou shalt not kill": Exodus 20:13, KJV.

78 "He [Allah] loves those": *The Meaning of the Holy Qur'an*, rev. trans. and commentary by 'Abdullah Yūsuf 'Alī (Brentwood, MD: Amana Corporation, 1989), *sura* 2:222.

78 "When ye prepare for prayer": Ibid., *sura* 5:6.

80 "Potentiality is prior in time to actuality": St. Thomas Aquinas, *Of God and His Creatures*, Chapter XVI, trans. Joseph Rickaby (1258–64; London: Burns and Oates, 1905), p. 86.

81 "As we know that it was chiefly": John Calvin, *The Institutes of the Christian Religion*, Book 1, Chapter 16, trans. Henry Beveridge (Woodstock, Ont.: Devoted Publishing, 2016), p. 91.

81 "He is created of both": Ibn Qayyim, *Kitab al-tibyan fi aqsam al-qur'an* (Beirut: Mu'assasat al-risala, 1994). See "Islam and Science," http://poraislam.page.tl/Islam-and-Science.htm.

82 "The books of Scripture": *Dei Verbum*, article 11, promulgated by Pope Paul VI on November 18, 1965. See http://www.vatican.va/archive/hist_councils/ii_vatican_council/documents/vat-ii_const_19651118_dei-verbum_en.html.

Transcendence

83 "I remember the night": William James, *Varieties of Religious Experience* (1902; Bibliobazaar edition, 2007), p. 71.

84 a letter that the French novelist: Freud refers to the letter from Rolland in *Civilization and Its Discontents*, trans. James Strachey (1929; New York: W. W Norton, 1961), pp. 11–12. He then goes on to interpret Rolland's "oceanic feeling"

in terms of his own theory of the ego and the profound attachment of the infant to its mother's breast.

84 "the lofty idealism of his literary production": "The Nobel Prize in Literature 1915," at https://www.nobelprize.org /nobel_prizes/literature/laureates/1915/.

86 "Truth, then, or Beauty is not": *Asia Magazine* 31 (1931): 139, reprinted in Abraham Pais, *Einstein Lived Here* (Oxford: Oxford University Press, 1994), pp. 102–3.

Laws

88 "a divine possession": "Life of Marcellus" (ca. AD 105), in Plutarch, *The Parallel Lives,* Great Books of the Western World, vol. 14 (Chicago: Encyclopaedia Britannica, 1952), pp. 252–55.

89 "Any solid lighter [less dense] than a fluid": "On Floating Bodies" (ca. 250 BC), in *The Works of Archimedes,* ed. T. L. Heath (Cambridge, UK: Cambridge University Press, 1897), Book I, Prop 5.

Doctrine

97 a principle I call the Central Doctrine of Science: I first introduced the term "Central Doctrine of Science" and its definition in my article "Does God Exist?," *Salon,* October 2, 2011.

97 There are a couple of implicit assumptions: Regarding the assumptions inherent in the Central Doctrine of Science: An important idea in philosophy, called the Principle of Sufficient Reason and most associated with the name of Gottfried Wilhelm Leibniz (1646–1716), states that every-thing must have a "sufficient" reason or cause. Philosophers debate what constitutes sufficiency. Is a cause "sufficient"

completely by virtue of its existence, whether or not it is intelligible by us or provable by us? Does sufficiency require that we fully understand the cause? Can certain aspects of a cause, such as the parameters in a scientific theory, be accepted as givens? These are deep philosophical questions, and their applicability to the scientific enterprise is complex and subtle. The Central Doctrine of Science, in my statement and understanding of it, does not require that we explain the causes of the causes. Scientists can believe that the universe is lawful—and even further believe in the validity of particular laws such as Einstein's equations for gravity—without knowing or even having a means of knowing where those laws came from.

99   the magnetic strength of the electron: What I call the "magnetic strength" of the electron is technically the "electron anomaly" multiplied by 1,000. See Eduardo de Rafael, "Update of the Electron and Muon g-Factors," *Nuclear Physics Proceedings Supplement* 234 (2013): 193.

101   "Our present theories are of only limited": Steven Weinberg, *Dreams of a Final Theory* (New York: Pantheon, 1992), p. 6.

103   "It seems to be one of the fundamental features": Paul Dirac, "The Evolution of the Physicist's Picture of Nature," *Scientific American* (May 1963).

Motion

105   "tremendous need for, shall I say the word": Steven Naifeh and Gregory White Smith, *Van Gogh: The Life* (New York: Random House, 2011), p. 651.

106   In 1990, medical researchers published: K. Arenberg, L. F. Countryman, L. H. Bernstein, and G. E. Shambaugh, "Van Gogh Had Meniere's Disease and Not Epilepsy," *Journal of the American Medical Association* 264 (1990): 491–93.

106 "The gravity and stillness of your youth": Shakespeare, *Othello,* Act 2, Scene 3.

106 "smooth mind": Emily Dickinson, "The Difference Between Despair," poem 305, in *The Complete Poems of Emily Dickinson,* ed. Thomas H. Johnson (Boston: Little, Brown, 1960).

108 "At the appointed hour": Terrien, *Le National,* February 19, 1851.

113 "Since that bore Kleiner": Albert Einstein to Mileva Marić, December 17, 1901, in *Collected Papers of Albert Einstein,* vol. 1, trans. Anna Beck (Princeton, NJ: Princeton University Press, 1987), p. 187.

114 "My parents are very distressed": Einstein letter to Mileva, August 30, 1900, in *Collected Papers,* vol. 1, p. 148.

115 "Try and penetrate with our limited means": Entry for June 14, 1927, *The Diaries of Count Harry Kessler,* ed. Charles Kessler (New York: Grove, 2002), p. 322. I am indebted to Walter Isaacson and his wonderful biography of Einstein, *Einstein: His Life and Universe* (New York: Simon and Schuster, 2007), for some of the Einstein quotations here.

115 "I'm not an atheist": In "What Life Means to Einstein," interview of Einstein by George Sylvester Viereck, *Saturday Evening Post,* October 26, 1929.

117 "At first, I was deeply alarmed": Werner Heisenberg, *Physics and Beyond,* trans. Arnold J. Pomerans (New York: Harper and Row, 1971), pp. 60–61.

117 "As the clock went past midnight": James D. Watson, *The Double Helix* (New York: New American Library, 1968), p. 118.

Centeredness

120 A 2008 survey by the Pew Research Center: "Religious Beliefs and Practices," Pew Research Center, Religion & Public Life Project. June 1, 2008.

120   A similar survey by the National Opinion Research Center:
      "Beliefs About God Across Time and Countries," National
      Opinion Research Center, University of Chicago, 2008.

121   "God is a person": Alvin Plantinga, *Warranted Christian
      Belief* (Oxford: Oxford University Press, 2000), p. vii.

121   At the beginning of 2017: Recent observations of the NASA
      Spitzer Space Telescope are reported in "These Seven Alien
      Worlds Could Help Explain How Planets Form," *Nature,*
      February 22, 2017.

Death

124   "because I wished to live deliberately": Henry David Tho-
      reau, *Walden* (1854), Chapter 2.

125   "Such harmony is in immortal souls": Shakespeare, *The
      Merchant of Venice,* Act V, Scene 1.

126   during a Buddhist retreat in Wisconsin: The Buddhist
      retreat referred to was held at the Christine Center, in Wil-
      lard, Wisconsin.

133   The neuroscientist Antonio Damasio has defined: Antonio
      Damasio's ideas about consciousness can be found in sev-
      eral of his books, for example, *The Feeling of What Happens:
      Body and Emotion in the Making of Consciousness* (New York:
      Harcourt Brace, 1999).

135   "When leaving my doctor's surgery": Leo from Tasma-
      nia, in "In Our Own Words: Younger Onset Dementia," at
      https://fightdementia.org.au/files/20101027-Nat-YOD-In
      OurOwnWords.pdf.

136   "My interaction with our grandchildren": "Ted's Story," at
      http://www.alz.org/living_with_alzheimers_8929.asp.

138   "Since I have shown it": Lucretius, *De Rerum Natura,* Book 3,
      trans. and ed. W. H. D. Rouse and M. F. Smith (Cambridge,
      MA: Harvard University Press, 1982), p. 221.

138   "Therefore death is nothing to us": Ibid., Book 3, p. 253.

Certainty

141   Augustine's "City of God": Augustine says that the City of
      God contains "absolute certainty," *City of God* (AD 413–427),
      chapter 18, trans. Reverend Marcus Dods, *St. Augustine's
      City of God and Christian Doctrine,* ed. Philip Schaf (Grand
      Rapids, MI: William B. Eerdmans, 2005), p. 590. See the
      Christian Classics Ethereal Library website: http://www
      .ccel.org/ccel/schaff/npnf102.titlepage.html.

142   "of Him out of Whose mouth nothing false": Saint Augus-
      tine, *To Consentius, Against Lying* (*Contra Mendacio*) (AD 420),
      paragraph 36. See website: http://www.newadvent.org/
      fathers/1313.htm.

142   "pious, true, holy chastity": Ibid., paragraph 38.

145   "Soul-sick was I, and tormented": *Confessions of St. Augus-
      tine* (AD 397–401), Book VIII, Harvard Classics, vol. 7 (New
      York: P. F. Collier & Son, 1909), p. 139.

145   "Take up and read": Ibid., pp. 141–42.

145   "Not in rioting and drunkenness": Romans 13:13, 14, KJV.

146   Augustine had found Certainty: A great deal of bio-
      graphical material about Augustine was recorded at the
      time by his friend Possidius, Bishop of Calama, and can
      be found at http://www.tertullian.org/fathers/possidius
      _life_of_augustine_02_text.htm.

147   Researchers at the California Institute of Technology: Ming
      Hsu, Meghana Bhatt, Ralph Adolphs, Daniel Traell, and
      Colin F. Camerer, "Neural Systems Responding to Degrees
      of Uncertainty in Human Decision-Making," *Science* 9
      (December 2005): 1680.

149   "When subjectivity is truth": Søren Kierkegaard, *Concluding
      Unscientific Postscript to Philosophical Fragments,* ed. and trans.
      Alastari Hannay (1846; Cambridge UK: Cambridge Univer-
      sity Press, 2009), pp. 170–71.

149   "the assumption of absolute determinism": Max Planck, *A*

Survey of Physical Theory, trans. R. Jones and D. H. Williams (1925; New York: Dover, 1960), pp. 67–68.

150 "We ought then to regard": Pierre-Simon Laplace, *A Philosophical Essay on Probabilities,* trans. Frederick Wilson Truscott and Frederick Lincoln Emory (1795; New York: John Wiley & Sons, 1902), p. 4; available online at http://bayes .wustl.edu/Manual/laplace_A_philosophical_essay_on _probabilities.pdf.

151 unknown to the hexagrams of the I Ching: Western and Chinese traditions differ on ideas of causality and connection. Chance and context play a larger role in Chinese philosophy than do mechanism and cause. In his introduction to the I Ching, psychologist Carl Jung suggests that the Chinese view each moment as the product of millions of small chance coincidences happening around it simultaneously, including the psychic state of a human observer. According to China scholar Joseph Needham, sequential events are connected as part of an organic whole, rather than as isolated causes and effects as in the Western worldview. Instead of saying, as in Western thinking, that A as a cause produced B as an effect, the Chinese use a "biological" model in which A is an origin and B is an end. In other words, B is somehow contained in A or is an outgrowth of A; A evolves fluidly into B, where each is part of a whole and cannot be isolated and studied as a separate entity. By contrast, in Western science, the conceptual if not physical isolation of elementary particles and forces has been an essential part of their study. These differences in worldviews may be one of the reasons that modern science, based on a reductionist methodology, developed in the West but not in the East. See Joseph Needham, *Science and Civilisation in China,* vol. 2 of *History of Scientific Thought* (Cambridge, UK: Cambridge University Press, 1957).

152 "Unless I am mistaken": Augustine of Hippo, *On Free Will,*

Book III, iv. 9. trans. J. H. S. Burleigh (388–395; Louisville, KY: Westminster John Knox Press, 2006), pp. 176–77.

158 A number of thinkers deeper than I: I am thinking, for example, of British philosopher Jeremy Bentham and his idea of utilitarianism.

Origins

159 On Wednesday, February 11, 1931: Einstein's attitudes toward a nonstatic cosmology and his trip to the Mount Wilson Observatory on February 11, 1931, are described in great detail in historian Harry Nussbaumer's "Einstein's Conversion from His Static to an Expanding Universe," *The European Physical Journal H* 39 (2014): 37–62.

159 When a prominent Belgian scientist: The Belgian cosmologist was Georges Lemaître.

159 "abominable": In Georges Lemaître's remembrance of talking to Einstein at the 1927 Solvay conference, "Rencontres avec A. Einstein," *Revue des Questions Scientifiques* 129 (1958).

159 "has smashed my old construction like a hammer blow": *New York Times*, February 12, 1931, p. 15.

160 "art the Creator of all times": *Confessions of St. Augustine*, Book XI, trans. Edward Bouvere Pusey (Religious Imprints, 2012), pp. 160–61.

161 Here's what we see in a time-lapse video: A good video showing cell divisions through a microscope is from the *Atlas de Reproducción Asistida*, and can be seen on YouTube at https://www.youtube.com/watch?v=P1h611sNji8.

164 "Then were created the gods": Alexander Heidel, *The Babylonian Genesis,* Tablet 1 (Chicago: University of Chicago Press), p. 18.

164 "If in the beginning": Rig Veda, translated in *Creation Myths of the World,* ed. David Adams Leeming (Santa Barbara, CA: ABC-CLIO, 2010), p. 3.

165 "The Originator is He": Qur'an, 2:117, translated in *The Message of the Quran,* by Muhammad Asad. See https://quran.com/2:117.

168 "relieves us of the necessity": R. H. Dicke, P. J. E. Peebles, P. G. Roll, and D. T. Wilkinson, "Cosmic Black-Body Radiation," *Astrophysical Journal Letters* 142 (1965): 415.

171 "the universe was neither created nor destroyed": Stephen Hawking, *A Brief History of Time* (New York: Bantam Books, 1988), p. 136.

171 "No cause is required": AL interview with Alex Vilenkin, July 7, 2015.

173 "What place, then, for a creator?": Hawking, *A Brief History of Time,* p. 141.

174 "As a Christian, I think there is a Being": AL interview with Don Page, September 11, 2015.

174 "One might think that adding": *The Preposterous Universe,* Sean Carroll's blog, March 20, 2015 (http://www.preposterous universe.com/blog/).

174 "In everyday life we talk": AL interview with Sean Carroll, August 4, 2015.

Multiverse

184 "benevolent influence": See Teun Tieleman, "Religion and Therapy in Galen," in *Religion and Illness,* ed. Annette Weissenrieder and Gregor Etzelmüller (Eugene, OR: Cascade Books, 2016).

Humans

191 "We procure means of seeing objects": Francis Bacon, *The New Atlantis* (1627), Harvard Classics, vol. 3 (New York: P. F. Collier & Son, 1909), pp. 177–78.

191 microphones able to hear: The MO-64 microphone made by Sanken in Japan. See http://www.sanken-mic.com/en /company/product.cfm.

194 "Science is as yet in its infancy": J. B. S. Haldane, "Daedalus, or Science and the Future," a paper read to the Heretics, Cambridge, on February 4, 1923. See https://www.marxists .org/archive/haldane/works/1920s/daedalus.htm.

194 A recent Gallup survey: "In U.S., 42% Believe Creationist View of Human Origins," at http://www.gallup.com /poll/170822/believe-creationist-view-human-origins.aspx.

195 "God made man in his own image": Genesis 1:28–31, KJV.

197 "O mankind!": *Sura* 4, Al Nisā', in *The Meaning of The Holy Qu'ran*, trans. and commentary by 'Abdullah Yūsuf 'Alī (Brentwood, MD: Amana Corporation, 1991), p. 183.

198 "The direct creation of Adam": Wahhab al-Turayri, "Biological Evolution, an Islamic Perspective," *Islam Today*, September 22, 2005 (http://en.islamtoday.net/artshow-416-2992 .htm).

203 "Not to go on all-fours": H. G. Wells, *The Island of Dr. Moreau* (1896), chapter 12. See, for example, *H. G. Wells Science Fiction Treasury* (New York: Chatham River Press, 1979), p. 105.

204 "Open the Air Lock": *Star Trek II: The Wrath of Khan* (1982), Paramount Pictures, written by Harve Bennett, with participating writers Jack B. Sowards and Samuel A. Peeples, The script can be found at http://www.imsdb.com/scripts /Star-Trek-II-The-Wrath-of-Khan.html.

205 "the lower animals, like man": Charles Darwin, *The Descent of Man, and Selection in Relation to Sex* (1871; Princeton, NJ: Princeton University Press, 1981), p. 39.

205 macaques refused to pull a chain: S. Wechkin, J. H. Masserman, and W. Terris, "Shock to a Conspecific as an Aversive Stimulus," *Psychonomic Science* 1 (1964): 47–48.

About the Author

Alan Lightman—who worked for two decades as a theoretical physicist—is the author of six novels, including the international best seller *Einstein's Dreams*, as well as *The Diagnosis*, a finalist for the National Book Award. He is also the author of a memoir, three collections of essays, and several books on science. His work has appeared in *The Atlantic, Granta, Harper's Magazine, The New Yorker, The New York Review of Books, Salon,* and *Nature,* among other publications. He has taught at Harvard and at MIT, where he was the first person to receive a dual faculty appointment in science and the humanities. Lightman is currently professor of the practice of the humanities at MIT. He lives in the Boston area.

A Note on the Type

This book was set in Monotype Dante, a typeface designed by Giovanni Mardersteig (1892–1977). Modeled on the Aldine type used for Pietro Cardinal Bembo's treatise *De Aetna* in 1495, Dante is a modern interpretration of the venerable face.

Composed by North Market Street Graphics,
Lancaster, Pennsylvania

Printed and bound by Berryville Graphics,
Berryville, Virginia

Designed by M. Kristen Bearse